Journal of Critical Infrastructure Policy
Volume 3, Number 1 • Spring/Summer 2022

Also from Westphalia Press
westphaliapress.org

The Idea of the Digital University

Dialogue in the Roman-Greco World

The History of Photography

International or Local Ownership?: Security Sector Development in Post-Independent Kosovo

Lankes, His Woodcut Bookplates

Opportunity and Horatio Alger

The Role of Theory in Policy Analysis

The Little Confectioner

Non-Profit Organizations and Disaster

The Idea of Neoliberalism: The Emperor Has Threadbare Contemporary Clothes

Social Satire and the Modern Novel

Ukraine vs. Russia: Revolution, Democracy and War: Selected Articles and Blogs, 2010-2016

James Martineau and Rebuilding Theology

A Strategy for Implementing the Reconciliation Process

Issues in Maritime Cyber Security

A Different Dimension: Reflections on the History of Transpersonal Thought

Iran: Who Is Really In Charge?

Contracting, Logistics, Reverse Logistics: The Project, Program and Portfolio Approach

Unworkable Conservatism: Small Government, Freemarkets, and Impracticality

Springfield: The Novel

Lariats and Lassos

Ongoing Issues in Georgian Policy and Public Administration

Growing Inequality: Bridging Complex Systems, Population Health and Health Disparities

Designing, Adapting, Strategizing in Online Education

Pacific Hurtgen: The American Army in Northern Luzon, 1945

Natural Gas as an Instrument of Russian State Power

New Frontiers in Criminology

Feeding the Global South

Beijing Express: How to Understand New China

The Rise of the Book Plate: An Exemplative of the Art

Journal of Critical Infrastructure Policy

Volume 3, Number 1 • Spring/Summer 2022

Richard M. Krieg, PhD, editor-in-chief

Westphalia Press
An imprint of Policy Studies Organization

Journal of Critical Infrastructure Policy
Volume 3, Number 1 • Spring/Summer 2022

All Rights Reserved © 2022 by Policy Studies Organization

Westphalia Press
An imprint of Policy Studies Organization
1367 Connecticut Avenue NW
Washington, D.C. 20036
info@ipsonet.org

ISBN: 978-1-63723-715-1

Cover and interior design by Jeffrey Barnes
jbarnesbook.design

Daniel Gutierrez-Sandoval, Executive Director
PSO and Westphalia Press

Updated material and comments on this edition
can be found at the Westphalia Press website:
www.westphaliapress.org

Journal of Critical Infrastructure Policy
Volume 3, Number 1 • Spring / Summer 2022
© 2022 Policy Studies Organization

Richard M. Krieg, PhD, Editor-in-Chief
TABLE OF CONTENTS

Editor-in-Chief's Letter 1
Richard M. Krieg

InfraGard: On the Front Line of Critical Infrastructure Protection 7
Editor's Interview with InfraGard National's President Maureen O'Connell

Preserving Ukraine's Electric Grid During the Russian Invasion 15
Thomas S. Popik

Defense Energy Resilience and the Role of State Public Utility
Commissions 57
William McCurry, Lynn Costantini, Wilson Rickerson, Jonathan Monken, Erin Brousseau

Expanding Oregon's Vision for a Once-in-a-Generation Infrastructure
Investment 73
Robert Parker and John Tapogna

Winter Storm Uri: Resource Loss & Psychosocial Outcomes of Critical
Infrastructure Failure in Texas 83
Liesel A. Ritchie, Duane A. Gill, and Kathryn Hamilton

Atlas for a Warp Speed Future: Enhancing Usual Operating Modes of the
U.S. Government 103
Amanda Arnold

Strengthening the Security of Operational Technology: Understanding
Contemporary Bill of Materials 111
Arushi Arora, Virginia Wright, and Christina Garman

A Risk-Informed Community Framework for the Assessment of Chemical
Hazards 137
Curtis L. Smith, Kurt G. Vedros, Kenneth F. Martinez, and David G. Kuipers

Journal of Critical Infrastructure Policy

Editor-in-Chief

Richard Krieg, PhD

Associate Editors

Noah Dormady, PhD

Judith (Cooksey) Krieg, MD, MS

Camille Palmer, PhD

Thomas Sharkey, PhD

Editorial Board

Richard Krieg, PhD, *Dept of Political Science, Texas State University*

Bilal Ayyub, PhD, *Dept of Civil & Environmental Engineering, University of Maryland*

Morgan Bazilian. PhD, *Payne Institute for Public Policy, Colorado School of Mines*

Chuck Bean, MA, *Executive Director, Metropolitan Washington Council of Governments*

Stephen Cauffman, BS, *Infrastructure & Devpt Recovery, US Dept of Homeland Security*

Arrieta Chakos, MA, *Principal, Urban Resilience Strategies*

Henry Cooper, PhD, *Director, Foundation for Resilient Societies*

Madin Dey, PhD, *Dept of Agriculture, Texas State University*

Noah Dormady, PhD, *John Glenn College of Public Affairs, Ohio State University*

Willard Fields, PhD, *Dept of Political Science, Texas State University*

David Flanigan, PhD, *Applied Physics Laboratory, Johns Hopkins University*

Ronald Gibbs, MA, *Harris School of Public Policy, University of Chicago*

Mary Lasky, MA, CBCP, *Chair, InfraGard National Disaster Resilience Council*

Richard Little, PhD, *Keston Institute Public Finance & Infrastructure Policy, USC (former)*

Pamela Murray-Tuite, PhD, *Dept of Civil Engineering, Clemson University*

Sean O'Keefe, MPA, *Maxwell School of Citizenship & Public Affairs, Syracuse University*

Camille Palmer, PhD, *Dept of Nuclear Science & Engineering, Oregon State University*

Edward Pohl, PhD, *Industrial Engineering Dept, University of Arkansas*

Kevin Quigley, PhD, *MacEachern Inst for Public Policy & Governance, Dalhousie University*

(cont'd.)

Craig Rieger, PhD, *Idaho National Laboratory*

Liesel Ritchie, PhD, *Dept of Sociology & Center of Coastal Studies, Virginia Tech*

Monica Schoch-Spana, PhD, *Center for Health Security, Johns Hopkins University*

Thomas Sharkey, PhD, *Dept of Industrial Engineering, Clemson University*

Chuck Wemple, *Executive Director, Houston-Galveston Area Council of Governments*

Editor-in-Chief's Letter

Richard M. Krieg, PhD

As stated in our website, the *Journal* is focused on "addressing the security and resiliency threats faced by US critical infrastructure sectors and the corresponding resiliency challenges of jurisdictions that rely on these infrastructures . . . each sector is considered so vital that its incapacitation would have a debilitating effect on the country's security, economic viability, public health and safety or other adverse outcomes." These goals are consistent with those of the InfraGard National Members Alliance (INMA), whose mission is to protect "US critical infrastructure and the American people by cultivating communications, collaboration and engagement between the public and private sectors; the Alliance unites the knowledge base, work and resources of these stakeholders to mitigate threats to national security, improve resilience, and strengthen the foundation of American life."

I am pleased to announce a new partnership between JCIP and INMA. With 80,000-plus members representing the sixteen critical infrastructure sectors, the Alliance will help promote *Journal* readership among private and public sector professionals working in critical infrastructures, analysts and researchers, and many others having industry-specific knowledge of the nation's critical infrastructure. Correspondingly, consistent with our manuscript submission and review policies, JCIP will accept manuscripts from InfraGard's broad membership.

In this issue, my Editor-in-Chief's interview with **Maureen O'Connell**, INMA President, provides background on InfraGard's development, organization, and priorities. InfraGard has grown, commensurate with a substantial increase in the volume and types of threats facing all critical infrastructures. Ms. McConnell's extensive background as a Special Agent with the Federal Bureau of Investigation personifies INMA's core public/private sector partnership with the FBI.

The InfraGard structure is a masterful arrangement of organizational components designed to enhance communications and coordination vertically and horizontally and across critical infrastructure sectors, geographic areas, public and private organizations—and through other internal mechanisms. The whole is greater than the sum of the parts. All of the moving organizational pieces are synchronized with a strong sense of mission: to strengthen safety, security, and resiliency across the country. In addition to the advancement of national critical infrastructure priorities, 77 InfraGard Members Alliances (IMAs) each maintain a dynamic local focus, and many IMAs operate Sector Chief Programs or Cross Sector Councils as they assess the local threat posture and develop appropriate programs.

This issue's lead article, "Preserving Ukraine's Electric Grid During the Russian Invasion," is authored by **Thomas Popik**, a prominent grid security analyst and infrastructure thought leader at the national and international levels. It examines an unprecedented situation—Ukraine's wartime management of its electric infrastructure. In developing his assessment, Mr. Popik used publicly available information and conferred with experts on energy policy, cybersecurity, electric grid operations, and nuclear safety in the United States and Ukraine. While Russia's ultimate objectives may be indecipherable, it has incentives to preserve reliable operation of the Ukrainian grid. Both sides have shown restraint in attacking energy infrastructure. An apparent Russian objective is keeping hard-to-replace critical infrastructures intact, especially hydroelectric and nuclear plants. Ukraine's natural gas pipeline system depends on grid electricity for its centralized control. Russia profits from transmission of its natural gas through Ukraine's pipelines to Europe.

Importantly, the article suggests avenues to boost electric power resilience in Ukraine. The Ukrainian grid is now interconnected with the electric grids of thirty-five other countries in Europe.[1] Collapse of the Ukrainian grid would likely cause highly adverse consequences—in Ukraine, as well as in Europe and Russia. Among other recommendations, the author points to the urgency of western democracies supplying additional funding support and other targeted assistance to rapidly increase Ukrainian grid resilience.

In wartime, events develop quickly. Mr. Popik's article was written at point in time. It contains several scenarios for Ukraine's electric grid over coming months. At the time of publication, events will have already overtaken the research. One critical observation—that Russia intends to "capture" power from Europe's biggest nuclear plant, Zaporizhzhia and use it for its own purposes—is already coming to fruition. On July 6, after the article was drafted, Russia announced its intention to take over Zaporizhzhia nuclear plant's operations and disconnect it from the Ukrainian power grid in September.

William McCurry, Lynn Costantini, Wilson Rickerson, Jonathan Monken, and Erin Brousseau report that more than 98 percent of military installations in the nation depend on the civilian power grid to execute their military and national security missions. In "Defense Energy Resilience and the Role of State Public Utility Commissions," the authors address a number of issues pertinent to Public Utility Commissions (PUCs)[2] developing relationships with the Department of Defense (DoD). The purpose would be to encourage projects that enhance national security while providing resilience benefits to local communities proximate to military installations. Such partnerships are warranted since DoD is served by

---

1   A. Blaustein, "How Ukraine Unplugged from Russia and Joined Europe's Power Grid with Unprecedented Speed. Scientific American," March 23, 2022.

2   PUCs are state public service commissions that regulate the utilities that provide essential services such as energy, telecommunications, power, water, and transportation.

hundreds of different utilities of varying types across different electricity market structures. Investor-owned utilities (IOUs) serve over 300 major military and national security installations across the country. In most states, IOUs are regulated by the state PUCs. As such, PUCs play an implicit and vital connective role between utilities they regulate and the defense communities that these utilities serve.

Using case examples and other analysis, the authors advise that PUCs across the nation would benefit from a greater understanding of the DoD energy resilience policy landscape as well as the roles they could play in relationships intended to protect critical infrastructure and national security. Consonant with a white paper released by the National Association of Regulatory Utility Commissioners (NARUC), key steps and considerations are recommended to position PUCs to explore such engagement. Importantly, the article examines how the value of defense energy resilience could be approached in relation to state-of-the-art resilience valuation as well as probable impacts of defense electric infrastructure investment. The authors note that NARUC will be providing guidance and resources during 2022 to assist PUC decision-making on defense energy resilience issues.

Earlier this year, the Texas Department of Health Services upward adjusted its official mortality figures for the State's February 2021 power grid crisis. The official toll now stands at 246 deaths, with victims ranging from less than a year to over 100 years of age.[3] The impacts in Texas of the disaster are still being felt. In "Winter Storm Uri: Resource Loss and Psychosocial Outcomes of Critical Infrastructure Failure in Texas," **Liesel Ritchie, Duane Gill** and **Kathryn Hamilton** examine persisting psychosocial stress and other adverse outcomes associated with the Storm in a sample of 1,567 Texans residing in impacted geographic areas

The article is important on a number of counts. First, it reflects planning guidance promulgated by the National Institute of Standards and Technology (NIST) on the importance of both critical infrastructure and social resilience—with recognition that social resilience requires an understanding of social functions and dependencies linked to the built environment. Second, the article builds on an important body of post-disaster research conducted by the authors and others, focusing on damages associated with critical infrastructure failure. The current analysis provides new insights into the continuing social repercussions of severe critical infrastructure breakdowns, specific populations at risk, and the need for planners and policymakers to address both the causes and long-term impacts of these events. In my personal view, the resilience of critical social infrastructure is a topic warranting far more attention in the critical infrastructure literature.

Arushi Arora, Virginia Wright, and Christina Garman, in "Strengthening the Security of Operational Technology: Understanding Contemporary Bill of Materials," provide new insights on a relatively recent safeguard capable

---

3   Texas Tribune, January 2, 2022: https://www.texastribune.org/2022/01/02/texas-winter-storm-final-death-toll-246/

of protecting critical infrastructures. A digital Bill of Material (BoM) defines the components—the list of ingredients—that make up a particular product. With the convergence of software and hardware in modern devices and instruments, BoMs can provide an exacting list containing the makeup of equipment or other product software and hardware components.

BoM tools are becoming increasingly important across various government sectors, as evidenced by a recent US executive order on cybersecurity (NIST 2021). They are especially useful for the enormous number of organizations reliant on international supply chains. These extensive supply chain operations can culminate in critical infrastructure components or instrumentation whose disruption can prove catastrophic. From the consumer perspective, a software BoM can enable verification of the software component sources, check for compliance with the consumer's policies, and scan for known vulnerabilities by performing software component analysis. It may also help in stimulating independent mitigations and risk-based decision-making related to software operation.

In addition to security and other purposes, BoMs can help prevent supply chain bottlenecks, ensure faster time to market, produce greater operational efficiency, and lessen the risk of errors. The authors provide classification suggestions for BoMs based on structure, functionality, component type, and architecture. They present case studies to highlight the security benefits of BoMs. In addition, the article identifies missing elements in BoM applications and suggests pathways for future research.

In "Atlas for a Warp Speed Future: Enhancing Usual Operating Modes of the U.S. Government," **Amanda Arnold** examines federal modes of governance that emerged in response to the COVID-19 pandemic. She notes that it is at the intersection of crisis modes of action and normal modes of operation where lessons have emerged from Operation Warp Speed[4] for consideration during non-crisis periods. Modes of action that arose during the pandemic break out into three categories: Speed, Scale and Scope. The article assesses each of these areas in the context of standard operating procedures, policy precedent, likely implications, and other factors. The mechanisms employed during the pandemic include flexible contracting authority, e.g. Other Transaction Authority (OTA), Emergency Use Authorization (EUA), and expanding the federal funding infrastructure—which typically involves a linear progression of research innovation—to one where steps were funded in parallel. An important factor was that vaccines were not being *invented* but were being *developed* following many years of research.

The insightful, policy-grounded article points to a need to devise principled, novel funding mechanisms for use in normal times that can flex to accommodate crisis speeds. In the case of medical countermeasure development, this

---

4 Operation Warp Speed was the federal effort that supported multiple COVID-19 vaccine candidates to speed up development.

means perfecting mechanisms that can operate in a manner that is appropriate to a crisis or a non-crisis situation, while ensuring elasticity, transparency and robust operation at the appropriate speed, scale, and scope. To that end, Ms. Arnold provides specific recommendations to adapt and expand the scope of the federal research and development infrastructure.

Robert Parker and John Tapagna, in "Expanding Oregon's Vision for a Once-in-a-Generation Infrastructure Investment," examine how the $864 billion Infrastructure Investment and Jobs Act (IIJA), passed in November 2021, could be adapted to long-standing state and local planning objectives. Using the Oregon case, where more than $5 billion in IIJA funding is anticipated, a lack of infrastructure funding has proven to be an insurmountable final barrier to long-standing state economic development plans. Aligning IIJA priorities with historic statewide planning goals would unlock Oregon's economic potential and provide compound dividends. To make the most of IIJA funding, the authors suggest that the Infrastructure Cabinet convened by Governor Kate Brown—and other top elected and agency officials—should focus part the State's IIJA outlay into a broad context of statewide economic recovery. The ultimate goal is to use these funds to stimulate multi-generational economic growth in Oregon.

In making its case, the article examines the State's highly prescribed land use and infrastructure planning processes. At present, there is a sizeable backlog of well-planned and verifiable development opportunities statewide. The authors explain how targeted IIJA investment in this area could be a primary determinant of economic development. Putting it all together, the authors formulate an agenda including positioning Oregon to grow its manufacturing sector, harness strategic infrastructure investment to allay the State's housing crisis and rebuilding and reconnecting Oregon's economically isolated neighborhoods and communities.

The United States' Chemical Critical Infrastructure Sector is immense, providing over $220 billion to GDP and accounting for one-fifth of world chemical production. Given the sector's magnitude, as well as the prospect of chemical terrorism, it is essential that communities gauge their risk and formulate responses to chemical release incidents. In "A Risk-Informed Community Framework for the Assessment of Chemical Hazards," **Curtis Smith, Kurt Vedros, Kenneth Martinez, and David Kuipers** present DRICAT (Data and Risk-Informed Chemical Assessment Technique). The process is relevant to local jurisdictions with little experience assessing chemical hazards and those with advanced programs wanting to benchmark and upgrade their current chemical incidents plans. DRICAT was designed to be reproducible, evidence-based, practical, and scalable for divergent communities and their unique potential chemical hazards.

InfraGard: On the Front Line of Critical Infrastructure Protection

Editor-in-Chief's Interview with Maureen O'Connell, President of the InfraGard National Members Alliance

Maureen O'Connell oversees the largest partnership committed to security of the nation's 16 critical infrastructure sectors. This is achieved through the mobilization of a huge membership (1 in every 4,000 Americans is an InfraGard member) that contributes private sector, industry-specific expertise to infrastructure and national security. With extensive background at the Federal Bureau of Investigation, Ms. O'Connell exemplifies the partnership that she leads. Since InfraGard's inception 25 years ago, the critical infrastructure landscape has dramatically changed. Each critical infrastructure sector is significantly more complex than was the case then. And 9/11, which occurred 5 years after InfraGard formation, heralded a period of rapid growth in the types and frequency of critical infrastructure threats. She was interviewed by JCIP Editor-in-Chief Richard Krieg in May, 2022.

Krieg InfraGard is a partnership between the Federal Bureau of Investigation and members of the private sector for the protection of U.S. critical infrastructure. The InfraGard program provides a vehicle for seamless public-private collaboration with government that expedites the timely exchange of information and promotes mutual learning opportunities relevant to the protection of critical infrastructure. Could you describe the organization's history and overall priorities?

O'Connell The InfraGard program celebrated its 25th anniversary last year and today represents the FBI's largest public/private partnership with over 80,000 members nationwide. The program's beginnings date back to

1996, when the FBI's Cleveland Field Office engaged experts from private industry for a cybercrime investigation. But in the 25 years, and especially since the events of 9/11, it's been proven that by working together, the FBI and the American business community can multiply their respective efforts to mitigate acts of crime and terrorism.

Eighty-five percent of U.S. critical infrastructure is owned by the private sector, which is why it's so beneficial for the FBI to engage with the American business community. The InfraGard program takes an all threats, all-hazards approach to defending our nation's most critical assets, with a focus on the 16 critical infrastructure sectors established by Presidential Policy Directive 21 (PPD-21). It is our goal to align with the FBI's threat priorities, including terrorism, cybercrime, insider threats, violent crime, fraud and much more.

Krieg You served as an FBI Special Agent for 25 years. How did your experience with the Bureau help shape your perspective as President of InfraGard National Members Alliance (INMA)?

O'Connell My career as an FBI special agent was one of the most meaningful and rewarding experiences of my life. I had the privilege of working alongside the most dedicated men and women in the world. Working for the FBI, and in law enforcement in general, absolutely shapes my perspective as President of InfraGard National Members Alliance. Our job every day was to investigate and bring justice to those who would do harm to our country or the American people. In my daily work, I often saw the worst of humanity, from the types of crimes being committed to the impacts on the victims.

I also worked specifically on the InfraGard program while assigned to the FBI Los Angeles Field Office as a Private Sector Coordinator. Private Sector Coordinators are responsible for maintaining an understanding of the FBI's engagement with private industry at the field office level and connecting them with the right FBI personnel to address whatever challenges they are facing. All of these experiences strengthened my resolve to continue being part of the solution even after I retired from the FBI. I am a very proud American with a lot of love for this country, and InfraGard is an opportunity to continue giving back to a nation that's given so much to me.

Krieg With the FBI's interest in cyber counterterrorism, counterintelligence and so on - how does outside collaboration help?

O'Connell Since the majority of America's national security and economic infrastructure rests within the business community, invariably that's where you'll find many of the threats. If you look at some high-profile cyber-attacks and insider threat cases in recent history, the targets are often private sector companies. They hold the key to our nation's innovation and prosperity, and bad actors realize this. Safety and security must be a shared responsibility between private industry and law enforcement if we are going to stay ahead of the threat.

The InfraGard program is such a powerful vehicle because it promotes ongoing dialogue and timely communication between its members and the FBI. This two-way exchange of information equips InfraGard members with the knowledge, information, and resources to protect their respective organizations - while the FBI benefits from private sector engagement, insight and expertise that can help prevent terrorism, cybercrime, espionage and more.

Krieg Let's turn to the organization itself. How is InfraGard structured? And how does the organizational structure help in building security and resiliency across each critical infrastructure sector?

O'Connell The private sector component of the InfraGard program is represented by InfraGard National Members Alliance, an FBI-affiliated independent nonprofit organization. INMA is comprised of 77 localized non-profit organizations called InfraGard Members Alliances (IMAs), and we represent the IMAs in relations with the FBI.

Each of the 77 local IMAs is affiliated with one of the FBI's 56 U.S. field offices, addressing the threats in their respective localities from coast-to-coast and border-to-border. This is important because the IMAs possess local and regional expertise yet remain a vital piece of the larger national security picture. InfraGard National Members Alliance, with our national focus, and the InfraGard Members Alliances, with their local focus, provide a complementary approach to strengthening safety, security, and resiliency across America.

InfraGard National Members Alliance also addresses sector-specific security with two of our flagship programs: the National Sector Security and Resiliency Program (NSSRP) and the National Cross-Sector Council Program (NCSCP). The NSSRP provides a vertical approach, contributing to the InfraGard mission through the creation and sustainment of a "network-of-networks" that fosters collaboration and information sharing among owners and operators of critical infrastructure within individual sectors. Led by Program Chair Dan Honore and a cadre of National Sector Chiefs, the NSSRP addresses the need for experts and intelligence for sector-specific activities.

Knowing there are many sector interdependencies and interoperabilities, the NCSCP provides a horizontal approach by addressing security threats and impacts that cross two or more critical infrastructure sectors. The NCSCP is led Program Chair Mary Lasky and seven National Cross-Sector Council chairpersons, providing the timely exchange of cross-sector specific information.

At the local level, many of the InfraGard Members Alliances have also built Sector Chief programs or Cross-Sector Councils to meet security and resiliency needs in their areas of responsibility.

Krieg Is it possible to say how your membership numbers break down by critical infrastructure sector?

O'Connell Currently, the Information Technology (IT) sector has the highest number of members in InfraGard, representing approximately 36 percent of our total membership. The Financial Services Sector represents about 10 percent of members, followed closely by Government Facilities (9 percent) and the Healthcare and Public Health Sectors (8 percent). Other prominent sectors include the Communications, Energy, Emergency Services, and Defense Industrial Base Sectors.

Krieg I'd like to focus on the organization's training and education functions – what are the principles that guide this work, how is it organized, and could you give examples of topics you are stressing?

O'Connell INMA strives to keep our fingers on the pulse of national threat priorities and provide content that's timely, relevant and valuable to our 80,000 plus members. National Infrastructure Security and Resilience U (NISRU) is our flagship eLearning platform, offering dozens of online courses, the Workshop Wednesdays series, and continuing education opportunities focused on critical infrastructure protection and resiliency. Through NISRU, we want to ensure that our members have the capabilities needed to meet the evolving landscape of critical infrastructure threats by providing comprehensive education, training, and workforce development for all 16 critical infrastructure sectors.

To provide an idea of the breadth of topics we cover, some recent NIS-RU offerings included "An Executive Approach to Cyber Risk Management", "Fundamentals of Homeland Security", and "Lessons Learned from Building the Intelligence Program at the NFL". Others include "Managing Risk in Supply Chain Software Applications", "Building an Effective Detection and Response Program", and "Tipping Point: Keys to Developing and Implementing a Comprehensive Violence Prevention Program."

Additionally, our National Sector Security and Resiliency Program (NSSRP) and National Cross-Sector Council Program (NCSCP) also produce webinars, events and information-sharing initiatives that are specific to individual sectors or cross two more sectors. For example, the National Disaster Resilience Council (NDRC), one of our Cross-Sector Councils, produces an annual summit. This year's edition in October will focus on the U.S. energy infrastructure and preparation for attacks and destructive natural events causing long-term power outages.

At the local level, InfraGard Members Alliances also produce numerous training and education programs that are customized to the security landscapes in their region. Their affiliations with local FBI Field Offices are vital in this regard, enabling the FBI to provide input and subject matter expertise pertaining to local threat priorities.

Krieg Given mounting concerns about cyber-attacks, what specific roles does InfraGard play to address that threat?

O'Connell Of the 16 critical infrastructure sectors, some are truly unique because of the enabling function they serve across all other sectors. The Information Technology Sector is one of them. America has become almost completely cyber-reliant, which has created our greatest Achilles heel. Cyber criminals, whether motivated by money or ideology, know this as well. That's why cyber represents the new battlefield.

InfraGard is based on collaboration, education, and information sharing, and this certainly holds true for addressing the cyber threat. Through our national programs, we work to provide educational webinars, workshops and courses on cyber threats and cybersecurity. Our FBI partners also collaborate with us to provide presentations and trainings, and we are very grateful for their insights and expertise. At the local level, InfraGard Members Alliances across the country are also sounding the alarm with cybersecurity-focused programs and events.

InfraGard members also gain access to the secure InfraGard Portal, which features the latest FBI intelligence, along with FBI and other government agency threat advisories, intelligence bulletins, vulnerability assessments and more.

Krieg At the time of this interview, the war in Ukraine looms large in the national mindset. Are there lessons for U.S. critical infrastructure resilience in that war?

O'Connell Initial reports indicate that Russia mounted several significant cyberattacks on Ukraine during the early phases of its offensive operations. Recent reports by Ukrainian officials and Microsoft have shown that most successes were of little strategic value to the Russian campaign. In fact, Ukrainian defenders were able to intercept an attack on the country's power grid - one that could have had significant consequences had it succeeded. The owners and operators of the U.S. critical infrastructure, across all sectors, should look at lessons learned from this conflict and prepare their own defensive strategies to counter recently demonstrated Russian tactics. Additionally, they should not believe that the worst is over. Experts agree that Russia has not unleashed its full cyber capability. CI owners and operators in the U.S. should seek out the latest information from sources such as CISA, and the FBI, as well as their software and systems integration providers. Information sharing between the public and private sector as well as private to private sectors will continue to be key to our collective defense against continued attacks from Russia, China, Iran - and other actors who are seeking to weaken our country through persistent cyberattacks.

Krieg As you look across the board in Infrastructure protection and resilience, what areas do you think most need attention in the next 5 years?

O'Connell One area needing attention would be to inspire our youth to rise to the challenge of becoming cyber experts. There are millions of high paying positions that sit vacant due to the low number of people with the requisite skills to fill them. We are at a critical point in this country and we all must work together to fill this void. InfraGard has a program called Cyber Camp in which we bring together young people of various ages to learn about cyber by interacting with private sector entities and FBI Agents in an interesting way. We have them work through challenges in both a business and law enforcement environment and get them excited about a career in cyber. Through problem solving together, participants also enjoy learning about teamwork.

We are also focused on the threat that China presents in every aspect of critical infrastructure—from supply chain issues to theft of intellectual property, to the deep entanglement of our respective economies. These issues represent a multi-pronged attack on our country and cause staggering financial damages to our businesses. The FBI Office of Private Sector created a 30-minute movie that provides context for this threat (note: the referenced video is footnoted below).[1]

Another focal point is the safety and security of our citizens, especially our most vulnerable, so school safety is a priority. Through our trainings mentioned above, and our associations with the nation's 80 strategically located fusion centers, we provide a holistic approach to school safety, including Security Assessments and recommendations. Many of the tools we provide can be used across all sectors. We provide Open-Source Intelligence (OSINT) training to teach our members how to identify red flags, and address threats, while providing clear instruction on how and where to report tips and leads. These things combined raise the overall security posture of the entity, thereby increasing safety.

Raising money for these programs is also a priority and can be done either as a direct donation or a sponsorship. We invite people to be part of the solution, and all donations are tax deductible (note: the referenced site is footnoted below).[2]

Krieg Finally, how would a subject matter expert involved in critical infrastructure—whether it be as a business executive, professional, law enforcement, military, attorney, etc.—join InfraGard, and what opportunities exist for participation?

O'Connell Joining InfraGard is one of the best opportunities to take an active role in strengthening the safety and security of your hometown—where it all starts—and working alongside your fellow Americans to help build a safe, secure, and resilient nation.

Membership is free. Applicants must be U.S. citizens who are employed or formerly employed within critical infrastructure for at least

---

1   https://www.youtube.com/watch?v=GdapE82GceA
2   https://inma.salsalabs.org/nonmemberdonate/index.html

three years. They must consent to a security risk assessment, which is unique to the InfraGard program and aims to create a higher level of trust among our membership. People who want to apply should visit www.infragard.org. On being accepted for membership, you'll join one of the local InfraGard Members Alliances and have access to their education and information sharing programs. For more information about InfraGard National Members Alliance, you can go to our web site at www.infragardnational.org.

The benefits of membership are vast and include but are not limited to 24/7 access to InfraGard's secure web portal; FBI and DHS threat advisories and alerts, intelligence bulletins, and analytical reports; unique networking opportunities; FBI and other government agency briefings and resources; virtual and in-person training and education events and programs produced by the FBI, INMA and InfraGard Members Alliances. And, as previously mentioned, eLearning courses and workshops presented by INMA's National Infrastructure Security and Resilience U (NISRU).

Most importantly, InfraGard members know they are truly making a difference where it matters. We want to create a safe and secure nation for all Americans—and the future generations that will inherit what we leave behind.

Preserving Ukraine's Electric Grid During the Russian Invasion

Thomas S. Popik[1]

[1] President, Foundation for Resilient Societies, thomasp@resilientsocieties.org

Abstract

In February 2022, the Russian Federation invaded Ukraine, a country with an interconnected electric grid.[1] While most analysts have concentrated on the military dimensions of Russia's attack, the implications for energy infrastructure are both unprecedented and critically important. Russia has incentives to preserve reliable operation of Ukraine's electric grid. Both sides have shown restraint in attacking energy infrastructure. An apparent Russian objective is keeping hard-to-replace infrastructure intact, especially hydroelectric and nuclear plants. Ukraine's natural gas pipeline system depends on grid electricity for its centralized control. Russia profits from transmission of its natural gas through Ukraine's pipelines to Europe. All countries seek to avoid infrastructure accidents and human migrations that are disruptive to their own societies. Nonetheless, by export ban, naval blockade, and physical attack, Russia has disrupted fuel supplies for Ukraine's generating plants. Interconnected electric grids are vulnerable to cascading collapse after forced outages of generating plants, transmission system disruptions, and deliberate attacks. As winter approaches, fuel supplies for Ukraine's electric grid will be constrained and the possibility of grid collapse increases. If Ukraine's electric grid were to be inoperable for a prolonged period, the result could be widespread death by famine, disease, and, in winter, hypothermia. Nuclear reactor meltdowns and spent fuel pool fires could also result, with radiation release extending beyond Ukraine's borders. Millions of refugees would cross the borders of Poland, Russia, Belarus, and other regional neighbors. Ukraine should develop a robust plan for electric grid restoration, including asking the Ukrainian people for their assistance during emergencies. To this end, financial and other targeted support for Ukraine's electricity sector by European

---

1. Ukraine's electric grid is an Interconnected Power System, sometimes also referred to as an Integrated Power System (IPS). Ukraine uses the terminology "Unified Energy System" (UES). For easier understanding of this paper by non-technical readers, the term "electric grid" is used throughout to mean the UES of Ukraine.

and other allies may be essential to reduce the prospect of long-term grid collapse. Events in Ukraine have public policy lessons for all nations with electric grids vulnerable to cascading collapse and long-term outage.

Keywords: Ukraine, Russia, invasion, infrastructure, electric grid, collapse

Introduction

For centuries, the geography and natural resources of Ukraine have made it a target for conquest, a subject of forced famine, and a venue for mass casualty events. Historical and geographic context is vital to understand the significance of famine in the Ukrainian national ethos, the most recent Russian invasion, and the prospect of massive population loss if the country's electric grid were to collapse.

Since establishment of the medieval state of Kyivan Rus', Ukraine has had its own political identity, language, and culture. The current boundaries of Ukraine are situated between European Union (EU) countries and Moldova to the west, the Russian Federation to the east, Belarus to the north, and the Black Sea and Sea of Azov to the south. In the latter part of the 18th century, most of Ukrainian ethnographic territory became part of the Russian Empire. Ukraine had a brief period of independence from 1917-20 but again came under Russian domination as part of the newly formed Soviet Union. In 1991, Ukraine regained its independence after the breakup of the Soviet Union, leaving a large minority of ethnic Russians living within its territory, especially in the Crimean Peninsula in the south and Donbas region in the east.

Ukraine's population before the 2022 invasion was 41 million, with a Gross Domestic Product (GDP) of US$517 billion, and per capita income of US$12,400 (2020 estimates). Seventy percent of the population lived in urban areas. Approximately 78 percent of Ukraine's citizens have Ukrainian ethnicity and 17 percent have Russian ethnicity; most of the remainder have eastern European ethnicity.[2]

Ukraine has a substantial land area of 550,000 square kilometers, about the same area as the state of Texas in the United States. The country's fertile black earth and ample rainfall makes for some of the most productive farmland in the world, a natural advantage that has provided persistent motivation for political domination and military conquest by other European states.

Despite its immense food production capacity, Ukraine has experienced repeated famines in which significant percentages of its population perished. Soviet policy precipitated the Povolzhye famine of 1921-23 in which one million died and

---

2   Unless otherwise noted, statistics in this article referring to Ukraine exclude the occupied territory of Crimea.

the Holodomor of 1932-33 in which four million died. (Internet Encyclopedia of Ukraine, 2022) Ukraine suffered 1.5 million civilian deaths due to famine and disease during World War II, many in urban areas. (World Population Review, 2022) In 1946-1947 Ukraine was hit with another famine, with 350,000 registered excess deaths. (Elman, 2000)

Figure 1. Map of Ukraine. (CIA Factbook, 2022)

The climate of Ukraine is temperate, but temperatures can fall to -20 degrees Centigrade (-4 degrees Fahrenheit) during the Siberian Anticyclone. Heating of homes and businesses is essential during winter months. Apartment buildings in Ukraine's cities rely on district heating systems constructed during the Soviet era. While some district heating systems are boiler-only, many are Combined Heat and Power (CHP) systems that use byproduct steam from electricity generation and industrial processes. Approximately 20 percent of heating in Ukraine is supplied by CHP plants. (Tsarenko, 2007)

Energy infrastructure has a prominent role in the history of Ukraine and the Soviet Union. Russia has the apparent objective of keeping energy infrastructure intact, especially Ukraine's four nuclear power plants, hydroelectric dams, and natural gas pipelines. Two major rivers, the Dnieper and the Dniester, have existing hydroelectric plants and significant undeveloped hydroelectric capacity. Ukraine's natural gas transmission system took decades to build and became an important energy source for Europe commencing with the Soyuz pipeline in 1978 and the Brotherhood pipeline in 1984.

Construction of the Dnieper Hydroelectric Station was one of the greatest infrastructure accomplishments of the early Soviet Union and part of the first

Five Year Plan. Its reservoir source, the Dnieper River, is the fourth largest river in Europe. The river runs north to south, entering Ukraine near the abandoned Chernobyl nuclear plant; flowing past the cities of Kyiv, Cherkasy, Dnipro, Zaporizhzhia, and Kherson; and exiting into the Black Sea. Original construction of the Dnieper Hydroelectric Station took place from 1927 to 1932, a period that overlapped the first stage of the Holodomor in 1932. Retreating Red Army troops dynamited the dam in 1941, causing between 20,000 and 100,000 Ukrainian civilians to be killed by the flood surge. The Soviet Union rebuilt the dam and hydroelectric facilities from 1944 to 1949 using turbines manufactured by General Electric. Installed capacity of this single hydroelectric plant is approximately ten percent of the peak demand of the Ukrainian electric grid post-invasion.

In February 2014, Russia invaded and then annexed the Crimean Peninsula. Shortly after the Crimea annexation, Russian separatists in Donbas began a proxy war with the Government of Ukraine. On February 24, 2022, the Russian Federation invaded Ukraine from Belarus in the north and Russia on the east with a military force of 100,000 troops.

Wartime disruption has already caused large population migrations. As of June 16, 7.7 million Ukrainians had evacuated across international borders. On June 18, the head of Russia's national defense, Mikhail Mizintsev, stated that 1.9 million Ukrainians had been "evacuated" to Russia since the invasion. Approximately 2 million Ukrainians evacuated early in the war but have since returned home after seeing that basic services, including supply of electricity, have continued to reliably operate.

Faced with an unprecedented challenge for the electricity sector, the Government of Ukraine and its utilities have been notably successful thus far. Operation of the electric grid during the invasion has been reliable, with no persistent outages except those caused by attack on local transmission and distribution facilities. Nonetheless, for reasons explained in this article, the Ukrainian electric grid is at risk of wide area, long-term collapse—especially as the next winter approaches. If such a collapse were to occur, infrastructure accidents, cross-border migrations, and mass casualties are likely outcomes. Prompt action by the Government of Ukraine with categorial financial support from western democracies can reduce the probability of long-term electric grid collapse.

Ukraine's Electric Grid and Supporting Infrastructures

While Ukraine's per-capita GDP is among the lowest in Europe, it possesses the infrastructure of an industrialized nation. Much of Ukraine's infrastructure, including its electric grid, was inherited from the Soviet era. Electricity infrastructure is oversized for the current demand and well past its design life. Since independence from the Soviet Union, Ukraine has spent significant societal resources in upgrading electricity infrastructure, including reducing carbon dioxide emissions through wind, solar, and biomass generation.

The electric grid is the keystone of Ukraine's infrastructure. Without a functioning electric grid, infrastructures in modern societies will fail, including those that provide services essential for human survival: water treatment and sanitation, food production and distribution, and winter heating. Without electricity for critical infrastructure, massive human casualties could result from famine, disease, and hypothermia.

The electric grid and its supporting infrastructures—telecommunications, nuclear, natural gas, coal, petroleum, railways, and seaports—are interdependent. Long-term collapse of the electric grid will cause failure of supporting infrastructures. In turn, degradation of supporting infrastructures puts the electric grid at risk.

Electric Grid

Ukraine's electric grid consists of electricity generating plants, high voltage transmission lines, transmission substations, distribution lines and substations, and customer loads. The transmission system operator, Ukrenergo, dispatches generating plants and controls the flow of electricity to distribution system operators. Because the electric grid is interconnected, all parts are synchronized, operating at the system frequency of 50 cycles per second (50 hertz).

Figure 2 shows major transmission lines of the Ukrainian electric grid operating at 220 kilovolts (kV), 330 kV, 400 kV, and 750 kV. In its current configuration, transmission lines connect the Ukraine grid to the systems of Poland, Slovakia, Hungary, Romania, and Moldova. A combination of long transmission lines and non-optimal power characteristics results in electricity losses of approximately 15 percent, a higher proportion than for most countries with interconnected electric grids. As a comparison, transmission and distribution losses in most European countries are 2-8%. (Council of European Energy Regulators, 2017)

Ukraine's electric grid has ample generation capacity, making it resilient to disruption and attack. To understand operation of Ukraine's grid, both installed capacity and actual generation are necessary statistics. Installed generating capacity of 55 gigawatts in 2020 had a wide diversity of energy sources: 25% nuclear, 51% fossil fuels (coal, natural gas, and oil), 11% hydroelectric, 9% solar photovoltaic, 2% wind, and 1% biomass. (State Statistics Service of Ukraine, 2020) Electricity generated in 2020 of 142,200 gigawatt-hours was 54% nuclear, 30% coal-fired, 8% natural gas-fired, 5% hydroelectric, 1% solar photovoltaic, and 1% wind.[3] (IEA, 2022) From 1990 to 2020, electricity generated in Ukraine fell by half. Because

---

3 Based on IEA data from IEA (2022) Electricity Information, IEA (2022), www.iea.org/statistics, All rights reserved; as modified by author.

much of Ukraine's generating capacity goes unused, the average capacity factor for 2020 was only 28%.

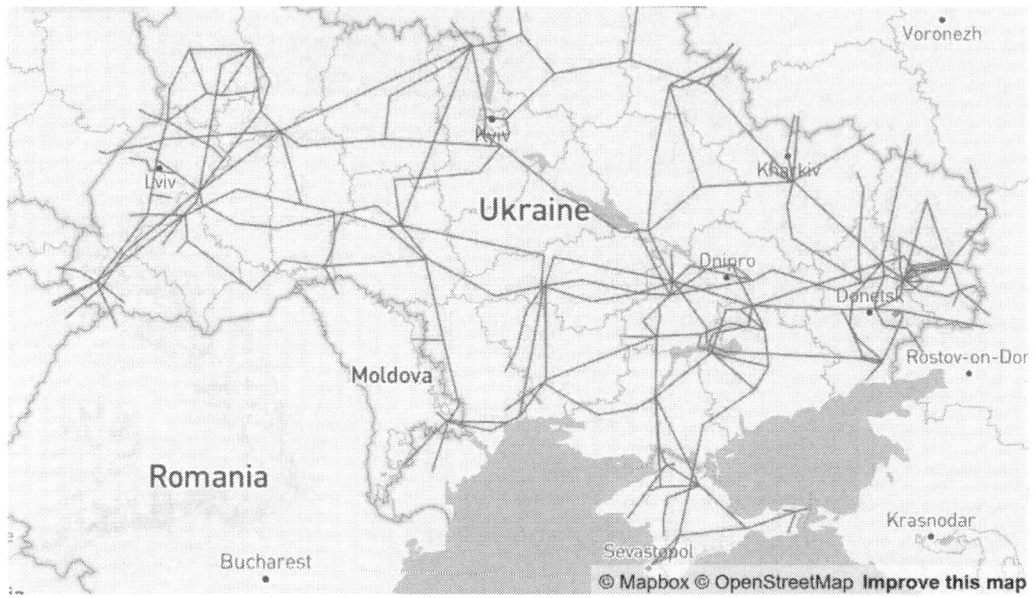

Figure 2. Major Transmission Lines of Ukraine's Electric Grid (Energy.info, 2022)

Electricity generation post-invasion has declined by 30-35% as compared to the same time the previous year. Installed capacity of dispatchable generation resources post-invasion is essentially the same as in 2020, although one-third of the capacity is now in captured territory and 5% has been destroyed. Nearly all wind resources are offline since the invasion. In regard to reserves of generating capacity, Ukraine's issue is not a lack of installed capacity, but the fuel required to reliably operate generating plants—particularly in winter months.

Ukraine's electricity exports were 4,754 gigawatt hours in 2020, representing average transmission of 543 megawatts. Current electricity export capacity to the ENTSO-E grid is 1,690 megawatts; with proper reactive power support, the same import capacity might be achieved.

Urban generating plants in Ukraine are commonly designed to supply by-product steam for district heating, so-called Combined Heat and Power plants. Eleven percent of Ukraine's generation capacity is CHP plants; before the invasion, these plants supplied 9% of Ukraine's electricity. (State Statistics Service of Ukraine, 2020) Ukraine's gas-fired CHP plants often have fuel oil as a backup energy source.

Nearly all of Ukraine's fossil fuel generation fleet is past its design life, making these facilities prone to breakdowns. The median commissioning year for Ukraine's thermal power plants is 1966; the median commissioning year for CHP plants is 1964.

Before the invasion, the Ukrainian grid was interconnected (and synchronized) with Russia's grid. On February 24, just a few hours before the invasion, the Ukrainian transmission system operator, Ukrenergo, disconnected from the Russian grid and initiated an "islanding" test.[4] Operation in island mode persisted until Ukraine interconnected with the ENTSO-E grid[5] on March 16, an acceleration of the prior 2023 integration timeline. Soviet-era transmission lines connecting Ukraine's system and former Eastern Bloc countries made this accelerated timeline possible.

Integration with ENTSO-E provides Ukraine with the opportunity to profit from export of electricity to European countries. On June 30, Prime Minister Denys Shmyhal announced 100 megawatts of exports to Romania, with potential exports of 2.5 gigawatts to Europe; this would provide state revenues of UAH 70 billion annually (US$ 2.4 billion). On July 1, the Ukrainian News Agency reported that Energoatom will supply 30% of Moldova's electric needs in July.

The integration of Ukraine's electric grid with ENTSO-E presents a new risk: a collapse of the Ukraine grid could cascade into the ENTSO-E grid. While settings on transmission system relays can prevent cascade, these settings could also reduce the resilience of the Ukraine grid to disruption and attack

Peak winter demand for Ukraine's electric grid before the invasion was 23.4 gigawatts, while average demand was approximately 16 gigawatts. (Ukrenergo, 2019) Peak demand has declined significantly since the invasion, with peak demand of 12 gigawatts for June 2022. (DixiGroup, 2022)

Transmission system operator Ukrenergo must manage daily fluctuations in electricity demand by maneuvering generating plants. Ukraine is a winter peaking system, with higher loads from heating systems during cold weather than from air conditioners during warm weather.

Before the invasion, the typical daily demand cycle in winter reached its low point in the early morning hours, rose to a morning peak around 10am, stayed nearly constant for much of the day, and rose slightly to a second peak around 6pm. Figure 3 shows the two daily load cycles in January 2019.

Ukraine has a deficit of generating plants that can maneuver to balance demand. The most maneuverable plants are the hydroelectric and pumped storage plants, but their flexibility is limited by seasonal river flows and run-of-the-river

---

[4] According to the Ukrainian transmission system operator, Ukrenergo, the February 24 islanding test had been planned well in advance. Therefore, the timing of the invasion just a few hours later on the same day appears to be coincidental. Nonetheless, the Russian Federation would have had a significant motivation to avoid military attack of a country whose grid was synchronized with its own, because a collapse of the Ukrainian grid could have cascaded into the Russian grid, with unpredictable consequences.

[5] European Network of Transmission System Operators for Electricity (ENTSO-E) is an association of transmission system operators (TSO) across Europe that maintain a frequency-synchronized electric grid. The ENTSO-E service area contains 39 TSOs in 35 countries across Europe.

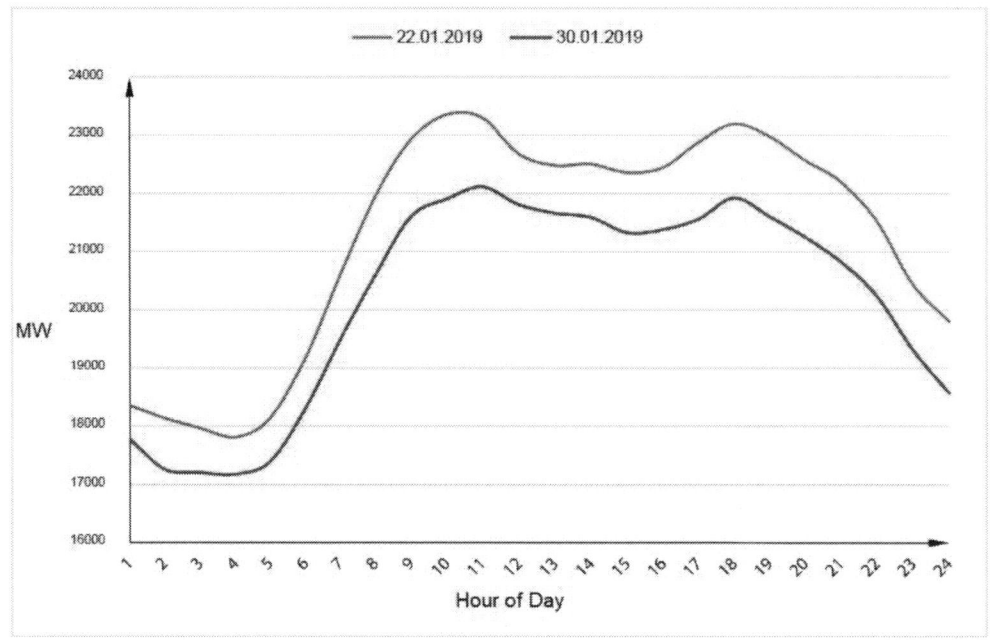

Figure 3. Graph of electricity consumption in the Interconnected Power System (IPC) of Ukraine on the working day of January 2019 at temperatures higher (1.5 ° C) and lower (-9.4 ° C) than the average monthly (-3.9 ° C) (Ukrenergo, 2019)

configurations at some plants. Seasonal river flows reduce the annual capacity factor of Ukraine's hydroelectric plants to approximately 20%, with flows peaking in springtime but low in winter. For 2020, the hydroelectric capacity factor was only 13%. (State Statistics Service of Ukraine, 2020) Ukraine has no significant capacity of fast-ramp aeroderivative turbines or combined-cycle gas turbines.[6] Steam turbines used in CHP plants are somewhat maneuverable, but in winter these plants must be run to match heating demand rather than electricity demand. As a result, Ukraine is forced to use smaller capacity coal-fired units to balance generation with demand. These coal-fired units are operated in start/stop mode over a 24-hour cycle, which causes equipment fatigue and increases the chance of forced outages.

Cyberattack is a threat to Ukraine's electric grid. A 2015 cyberattack on a distribution system operator caused a blackout for 230,000 consumers. The 2015 attack demonstrated that the Ukraine's electric grid is vulnerable to cyberattack, at least at the distribution level. (Zetter, 2015) Russia has apparently used Ukraine as a testing ground for cyberattack techniques that could be used on other electric grids.

---

[6] Aeroderivative turbines are based on the design of aircraft engines, running on natural gas. These turbines have low weight, high efficiency, and can respond quickly to changes in electricity demand. Combined cycle gas turbines (CCGT) combine first-stage gas turbines with second-stage steam turbines that use waste heat from the gas turbines.

There have been several attempts to cyberattack Ukraine's electric grid since the Russian invasion. The first attack reportedly began on March 19 shortly after integration of Ukraine's grid with ENTSO-E. (O'Neill, 2022) Another attack reportedly took place in early April. (Bajak, 2022). If successful, these attacks would have affected 2 million consumers, approximately 12% of Ukraine's 16.8 million household consumers and 579,000 non-household consumers.

Physical attacks on Ukraine's generating plants have been few and while deliberate, appear to be mostly incidental actions not designed to collapse the electric grid. Physical attacks on generating plants include Vuhlehirska Thermal Power Plant (TPP) at 3600 megawatts capacity; Kryvyi Rih TPP at 2,925 megawatts; Trypilska TPP at 1,825 megawatts; Sievierodonetsk TPP at 260 megawatts; Kremenchuk TPP at 255 megawatts; Chernihiv CHP at 210 megawatts; and Okhtyrka CHP at 13 megawatts. The attack on the Zaporizhzhia Nuclear Power Plant (NPP), with capacity of 6,000 megawatts, had the apparent objective of intact infrastructure capture; this plant has continued to operate and feed power into the transmission system. A significant number of Ukraine's generating plants have been captured by Russia or are close to captured territory. Figure 4 shows the locations of Ukraine's dispatchable generating plants[7] with capacity of 100 megawatts or more as of June 30, 2022.

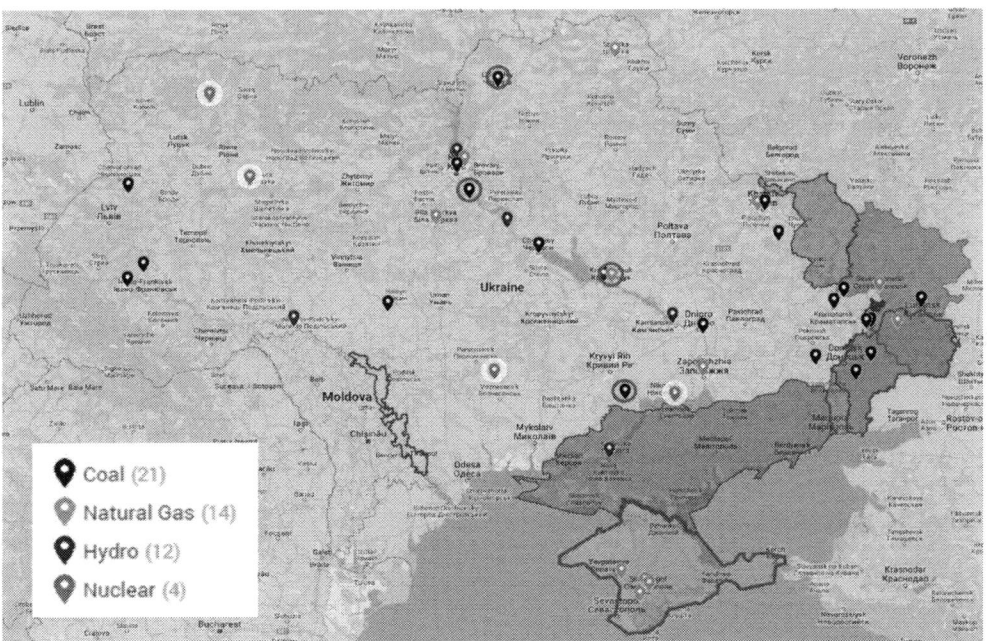

Figure 4. Dispatchable Generation Plants in Ukraine with Capacity of 100 megawatts or Greater. Territory captured by Russia as of June 30, 2022 is shaded. Red circles indicate intentional attacks on plants that are not in captured territory. (Google My Maps, 2022; Project Owl, 2022)

---

7   Dispatchable generating plants are those that can increase or decrease electricity generation on command.

Since the invasion, Ukraine's generating capacity has declined from attack damage, capture, and fuel shortages. On June 15, 2022 Ukraine's Minister of Energy Herman Halushchenko stated:

> Ukraine's energy infrastructure continues to suffer devastating damage. Almost 650,000 consumers do not have access to electricity, and more than 180,000 households do not have gas. Almost 5% of the installed generating capacity was destroyed. In addition, 35% of generating capacity is now in the occupied territories. Russian aggression has disrupted more than 50% of heat capacity, 30% of solar, and more than 90% of wind generation. Gas production has fallen by about 12-15%.

There have been no reported attacks on major transmission system nodes in Ukraine's electric grid. Russian missile attacks on rail traction substations on April 24-25 were on the distribution system. System disturbance from these rail traction station attacks blacked out portions of Lviv, but for less than a day.

Telecommunications

Reliable telecommunications are essential for transmission and distribution control of wide-area electric grids. Ukraine has a well-developed telecommunications infrastructure, with long-haul connectivity provided by state monopoly Ukrtelecom. Thus far in the invasion, service has been reliable, as evidenced by continued voice, internet, and video communications among Ukraine and other countries.

Since the invasion, two notable telecom service interruptions have occurred. On March 28, internet tracker NetBlocks reported a "massive cyberattack" on core telecommunications infrastructure, with connectivity across Ukraine collapsing to 13 percent of prewar levels. A Tweet by the State Service of Special Communication and Information Protection (SSSCIP) of Ukraine reported that the attack had been neutralized by 12:35pm. (Condon, 2022) Also according to SSSCIP, on April 30, internet broadband and cell phone service to the Kherson region was cut off. By May 4, internet service was restored. An analysis by Cloudflare, an American content delivery network, revealed that Russian occupiers had shifted routing of internet traffic from Kyiv data centers to Moscow data centers and then back to Kyiv, a probable cause of the outage. (Tome, 2022)

Nuclear

Ukraine has an extensive nuclear sector with 15 nuclear reactors at four plant locations, for a total of 13.8 gigawatts capacity. Before the invasion, half of Ukraine's electricity was generated by nuclear power plants. Ukraine's nuclear fleet is significantly oversized (or underutilized) with a capacity factor of 59% in 2020. During the invasion, eight reactors at four plants have continued to operate and feed power into the electric grid: two at the Zaporizhzhia NPP, three at the Rivne NPP, two at the South Ukraine NPP, and one at the Khmelnytskyi NPP. (International Atomic Energy Agency, 2022)

Figure 5. Location of Ukraine's Nuclear Power Reactors (Energoatom)

Ukraine's reactors have a Water-Water Energetic Reactor (VVER) design, similar to U.S. Pressurized Water Reactor (PWR) designs. Unlike the RBMK reactors at Chernobyl, VVER reactors have steel and concrete containment vessels. The safety systems of VVER reactors are similar to PWR designs. VVER reactors in Ukraine have backup diesel generators to allow safe shutdown and continued cooling when grid power is unavailable. (Nuclear Energy Institute, 2022).

On March 3, Russian forces captured the Zaporizhzhia NPP, the largest nuclear plant in Europe, having six reactors with a total capacity of 6,000 megawatts. During the attack, artillery shelling set an administrative building on fire and hit one of the containment vessels. Fortunately, no radiation release occurred. Russian forces have held the plant's operators captive since March 4. Rosatom, Russia's nuclear utility, has sent engineers to Zaporizhzhia NPP to monitor plant operations.

On a visit to occupied Ukraine, Russia's Deputy Prime Minister Marat Khusnullin threatened that Zaporizhzhia's electricity will be diverted to Russia if Ukraine does not pay for the power, saying, "If Ukraine's power system will be ready to pay, then we'll work; if it won't, the plant will work for Russia."

The owner of the Zaporizhzhia nuclear plant, Energoatom, has dismissed diversion of its generated electricity to Russia as a technical impossibility, because the plant is not connected to the Russian electric grid. (The Moscow Times, 2022) Moreover, the Ukraine and Russian grids are no longer synchronized.

While the dispute over the Zaporizhzhia nuclear plant's electricity may be wartime posturing, existing transmission infrastructure could make the Russian power diversion scheme eventually feasible, even if Russian territory capture is limited to the Donbas and southern Ukraine. The Volgograd–Donbass High Voltage Direct Current (HVDC) transmission line runs from the Mikhailovska converter station, situated northeast of Pervomaisk, to a terminal at Volga Hydroelectric Station in Russia. (Pervomaisk is a city with population of 38,000 in the self-proclaimed "People's Republic of Luhansk," an area of Ukraine controlled by Russian separatists.) This 475-kilometer line, originally designed to carry 750 megawatts, is currently operated at 100 megawatts. (Wikipedia, 2022) Renovations for the line are ongoing. Because AC current from the Ukrainian grid would be converted to DC for transmission, it could be technically possible to supply the Russian grid with power from Zaporizhzhia NPP. This potential scheme gives Russia incentive to keep the Zaporizhzhia nuclear plant operating, which may explain why Russia made it a priority to capture the plant in the first two weeks of the invasion.

Natural Gas

Natural gas is an important fuel for Ukraine's electric generating plants, especially CHP plants providing district heat to cities. Ukraine has significant natural gas resources with over 1 trillion cubic meters of reserves, the second largest in Europe after only Norway. These gas reserves provide a motive for Russian capture of Ukraine's territory. A small proportion of Ukraine's natural gas reserves is available for use during the war. Development of Ukraine's gas reserves requires unconventional technologies, such as horizontal drilling.

Ukraine has an extensive natural gas pipeline system inherited from the Soviet era. This system is used for transit of gas from Russia to Europe, as well as in-country distribution. The pipeline operator is Gas Transmission System Operator of Ukraine (GTSOU), a regulated monopoly. Before the February 2022 invasion, one-third of Europe's natural gas transited Ukraine.

Figure 6 shows the pipeline network transmitting gas from Russia to Europe. Gas from the Soyuz (capacity of 32 billion cubic meters (bcm)/year) and Brotherhood (28 bcm/year) pipelines transit Ukraine; additionally, a branch from the Yamal-Europe pipeline transits Ukraine. Other Russian natural gas pipelines supplying Europe include the Nord Stream 1 (59.2 bcm/year), the Yamal-Europe (33 bcm/year), and the TurkStream (31.5 bcm/year). Nord Stream 2 has double pipelines with capacity of 27.5 bcm/year each, but this facility has never gone into operation.

Taken together, shipments of Russian gas to Europe at the end of June 2022 have been running at about one-third of their pre-invasion levels. Sanctions have

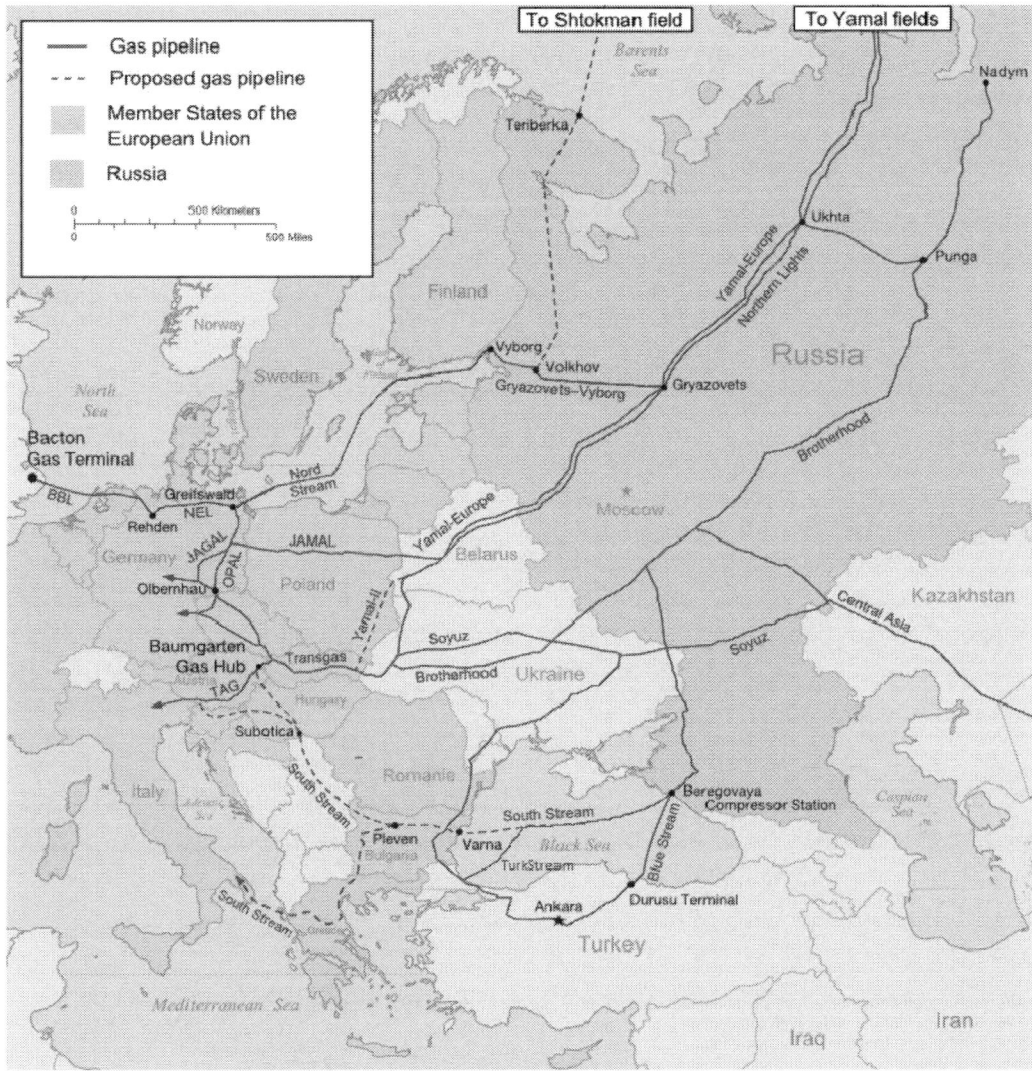

Figure 6. Major Natural Gas Pipelines of Europe.
(Wikimedia Commons, Samuel Baily, CC BY 3.0 NZ; 2022)

prevented use of the Nord Stream 2 pipeline since its completion in September 2021. On June 15, 2022, Russia reduced flow in the Nord Stream 1 pipeline to about 40% of capacity for compressor maintenance. Deliveries of gas to Europe through the Yamal pipeline had been intermittent; on May 12, Russian banned all shipments of gas through this pipeline due to a payment dispute with Poland; some flow has since resumed. The TurkStream pipeline has continued to operate except for maintenance during June 21-28; gas flow has since resumed.

On May 11, the Gas Transmission System Operator of Ukraine (GTSOU) shut down flow through the Sokhranivka interconnector station after interference by Russian forces; since then, flows of Russian gas through the remaining Sudzha interconnection station have been in the range of 40-45 million cubic meter (mcm)/day, down from the pre-invasion "contract volume" of 109 mcm/day. Ukraine's pipelines still carry about 30% of gas transit to Europe because of reductions in Nord Stream 1 flows and intermittent flows in the Yamal pipeline.

Ukraine's natural gas consumption for electricity generation, heating, and industrial use in 2021 was 26.1 bcm while production was 18.6 bcm. (British Petroleum, 2022) Gas to fill the past production gap of 7.5 bcm had been contractually supplied by European countries, although most gas imported by Ukraine is "resold" Russian gas. For 2022, the government estimated annual gas production of 16-19 bcm; estimated consumption for the year is 21-24 bcm. (Interfax, 2022)

Storage capacity for natural gas in Ukraine is 31 bcm, the largest in Europe and approximately one year's consumption. As of March 12, stored gas was 9.5 bcm, according to Ukraine Prime Minister Denys Shmygal. (Watts, 2021) As of June 1, stored gas was 10 bcm, with a Government of Ukraine goal of 19 bcm stored by the beginning of the next heating system.

Non-critical attacks on Ukraine's natural gas system have reduced supply. For example, on February 27, a Russian attack caused a pipeline explosion in Kharkiv. On March 13, a Russian attack severely damaged a gathering station for the Shebelinka gas field.

There have been no physical attacks on major compressor stations of Ukraine's natural gas system, but there has been Russian interference with contracted gas flow. On April 8, GTSOU reported gross interference by groups allied with Russia at the Novopskov compressor station in the Luhansk separatist region. Nearly one-third of Ukrainian gas transit to Europe had passed through this station. On May 10, GTSOU declared force majeure and refused to accept gas from the Novopskov compressor station for transport through its Sokhranivka pipeline. (GTSOU, 2022)

Coal

Along with natural gas, coal is an important fuel for Ukraine's electricity generation. Ukraine's coal reserves are the seventh largest in the world. There are three major coal basins in Ukraine: Lviv-Volyn with bituminous coal, Dnieper with lignite coal, and Donbas with bituminous and anthracite coal. The vast majority of Ukraine's coal reserves are in the Donbas region, the same area increasingly under Russian control.

Before the invasion, Ukraine depended on imports for approximately half of its coal consumption. (IAE, 2020). In the period from January to November 2021, Ukraine received 75% of its coal imports from Russia, 17% from Kazakhstan and 8% from the US. (The Coal Hub, 2021). In November 2021, Russia stopped exporting coal to Ukraine. Nor can Ukraine presently import coal through its seaports, as they have been blockaded by Russia.

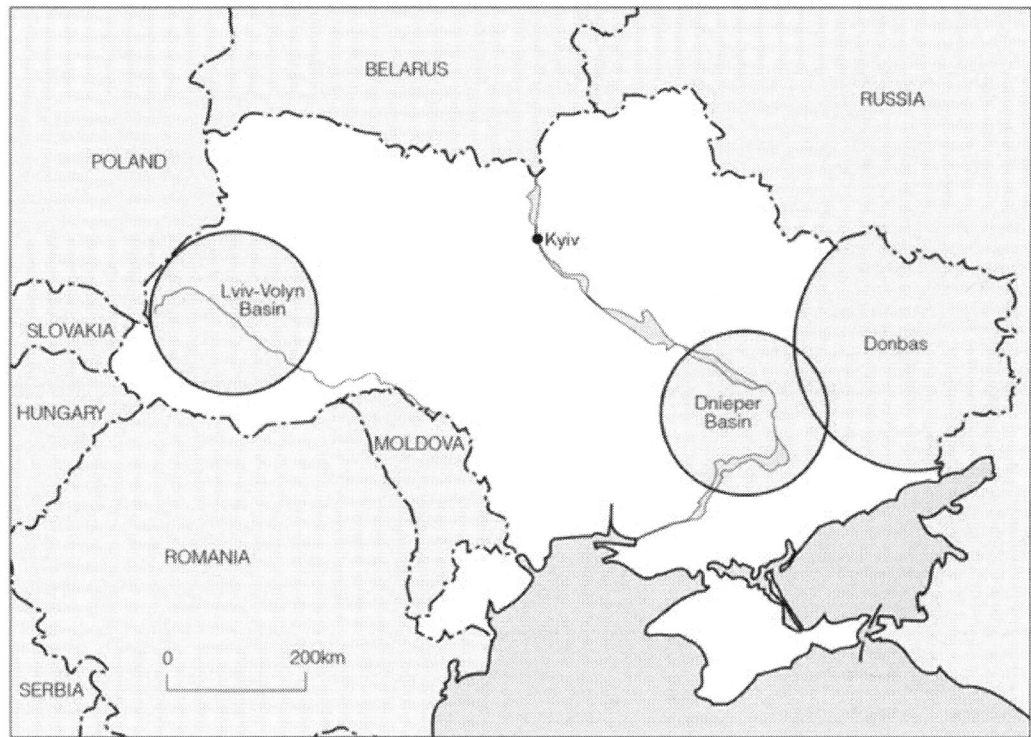

Figure 7. Coal Basins in Ukraine. Source: IEA/Oprisan (2011), "Prospects for coal and clean coal technologies in Ukraine," www.iea.org. All rights reserved. With author annotation for city of Kyiv.

Figure 8. Oil Pipelines in Ukraine. Source: IEA/Oprisan (2011), "Prospects for coal and clean coal technologies in Ukraine," www.iea.org. All rights reserved. With author annotations for Kremenchuk Refinery and Thermal Power Plant; Odesa-Kremenchuk pipeline.

Petroleum

Petroleum is a backup fuel for a significant number of Ukraine's generating plants, but supplies have been interrupted since the invasion. In 2021, Ukraine consumed 239,000 barrels per day of petroleum liquids while in-country refinery production was 73,000 barrels per day. (British Petroleum, 2022) Before the war, seventy percent of refined petroleum had been imported from Belarus and Russia, with most of the remainder supplied by in-country refineries. Crude oil had been refined in the Kremenchuk refinery with throughput of up to 68,000 barrels per day. Additionally, the Shebelynka gas plant has capacity for approximately one thousand barrels per day of refined petroleum.

The southern branch of Russia's Druzhba pipeline network passes through Ukraine and supplied crude oil to Slovakia, Hungary, and the Czech Republic before the invasion. Ukraine's pipeline operator, Ukrtransnafta, has a contract with Russia's Transneft to transport oil through 2030. Approximately 244,000 barrels per day of Russian crude oil traveled through Ukraine in 2020. (U.S. Energy Information Agency, 2022).

The bi-directional Odesa-Brody pipeline with a capacity of 70,000 barrels per day between Poland and the Yuzhne maritime terminal near Odesa had been an additional channel for oil transit. In some years, the pipeline had been used to transport Russian oil from Poland to Yuzhne. In 2020, the Odesa-Brody pipeline resumed the transit of oil from Yuzhne to Poland.

The logistics of petroleum import to Ukraine during the invasion have been challenging, especially because Ukraine and Poland's rail gauges are incompatible. Moreover, attacks on Ukraine's petroleum transport, refining, and storage have contributed to severe shortages. As of May 13, Russia had destroyed 27 fuel storage facilities in Ukraine. (Reuters, 2022) The Shebelynka gas plant was shut down on February 26, two days after the start of the war. On April 2, a missile strike on the Kremenchuk refinery destroyed nearby fuel storage, resulting in the refinery's deactivation. The same missile attack targeted Kremenchuk TPP. Also on April 2, Russian forces attacked fuel storage tanks at the Port of Odessa and an inactive refinery nearby. On April 24-25, Russia attacked rail traction stations in western Ukraine to impede import of petroleum. Also on April 24-25, Russia again attacked the Kremenchuk refinery and nearby thermal power plant.

During President Zelensky's April 29, 2022 evening address, he stated:

> *The occupiers are deliberately destroying the infrastructure for the production, supply, and storage of fuel…Russia has also blocked our ports, so there are no immediate solutions to replenish the deficit… But government officials promise that within a week, maximum two, a system of fuel supply to Ukraine will be at work that will prevent shortages.*

On May 13, Ukraine's Minister of Economy Yulia Svyridenko announced measures to increase refined fuel imports from the European Union via rail, truck, and pipeline. Imports were to be 12,000 tons/day or the equivalent of 82,000 barrels per day, approximately one-third of prewar consumption. (The Odessa Journal, 2022).

Railways

Railways are a major means of transporting coal for electricity generation. Ukrzaliznytsia, a state-owed monopoly, operates most of the Ukrainian rail system. Ukraine has 13,447 miles of track of which almost half—6,138 miles—is electrified. Because the rail sector was part of the Soviet system, it has a wider gauge than the EU rail system, necessitating car transloading terminals at borders. A single mountain passage, the Beskydy Tunnel, handles sixty percent of Ukraine-EU freight. Figure 9 shows the Ukrainian rail system and the locations of attacks.

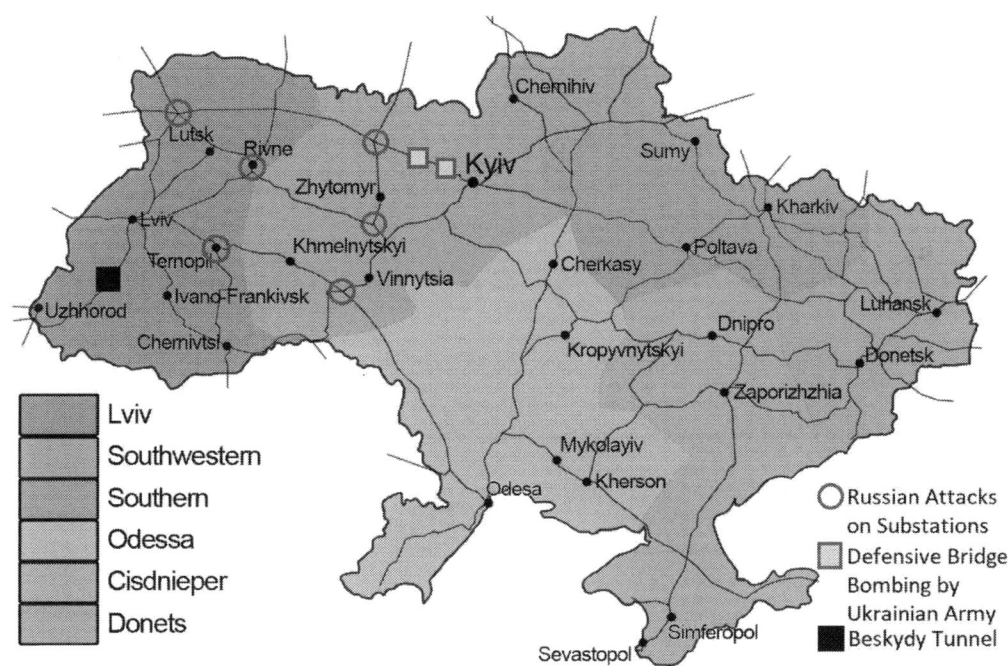

Figure 9. Attacks and Disablements of Ukraine Railway System as of April 25, 2022 (Wikimedia Commons, Terek, CC BY-SA 4.0; with author annotations.)

Until late April, there were few attacks on Ukraine's rail system and one of the attacks that did occur was directed at human evacuation rather than the transport of goods. On April 8, missiles hit the Kramatorsk station in eastern Ukraine, killing fifty of the crowd of 4,000 waiting for the first train of the day. Russian targeting of rail infrastructure escalated on April 24-25 when six rail traction substations in western Ukraine were hit by missiles, destroying transformers that convert high voltage electricity from transmission lines to lower voltages necessary for use by electric locomotives. This disruption to the distribution system caused a temporary blackout for parts of the city of Lviv. Also on April 24-25, Russian forces attacked a railroad bridge across the Dniester Estuary leading to the Danube ports; these ports are an alternative to ports on the Black Sea. Despite these attacks, Ukrainian railway operators have continued substantial transport of goods.

Seaports

Before the invasion, Ukraine's Black Sea ports imported coal and oil for electricity generation. Major ports include Berdvansk with 9 berths and a capacity of 2.1 million tons annually; Chornomorsk with 29 berths and capacity of 21.5 million tons; Kherson with 10 berths and a capacity of 8 million tons; Mariupol with 21 berths and capacity of 7.6 million tons; Mykolaiv with 23 berths and capacity of 24 million tons; Odesa with 46 berths and capacity of 31.4 million tons; and Yuzhnye with 38 berths and capacity of 61.7 million tons. Yuzhnye had been a major terminal for coal and oil imports.

During the early invasion, Russian forces captured the ports of Berdvansk, Mariupol, and Kherson. The port of Mykolaiv had been captured, but Ukrainian forces repulsed the invasion by mid-March. Russia attacked the Odesa petroleum storage facility by missile on April 3.

As of the writing of this article, over ninety percent of Ukrainian port capacity (excluding annexed Crimean ports) is still controlled by the Government of Ukraine. Nonetheless, all of Ukraine's ports are closed due to mining of the Black Sea and an ongoing Russian naval blockade. Coal and petroleum imports through seaports have halted. Even if the Russian blockade were to end, maritime insurance companies have been unwilling to write coverage for ship charters to Ukrainian ports.

Russian Incentives

From the Russian invasion of Crimea in April 2014 until the current invasion in February 2022, Ukraine had been in de facto war with Russia, especially in the Donbas region. Nonetheless, the Russian and Ukrainian electric grids were interconnected throughout this period, with daily coordination of cross-border transmission capacity. Uncoordinated transmission of power from Ukraine to Russia could have caused a cascading collapse of the Russian grid. Likewise, uncoordinated disconnection of the Russian grid from Ukraine's could have caused a cascading collapse in Ukraine. Although at war, Russia and Ukraine had shared incentives not only for reliable electric grid operation, but also to avoid attacking the opposing side's energy infrastructure.

On February 24, 2022, the Russia-Ukraine incentive structure changed. Russia invaded. Ukraine disconnected from the Russian grid for an "islanding test," just hours before the Russian invasion. The Ukraine electric grid is now synchronized with the ENTSO-E grid. Russia and Ukraine have established a new incentive structure for energy infrastructure.

Russian incentives in the present instance are difficult to definitively determine but may be implied by behavior. Apparent Russian incentives include maintenance of mutual restraint on attacking energy infrastructure, keeping energy infrastructure intact, profits from transmission of natural gas to Europe, prevention of infrastructure accidents, and avoidance of disruptive human migrations. While the wartime goals of Russia and Ukraine are certainly opposed, Ukraine can nonetheless consider how Russian incentives may be consistent with its own interests.

Mutual Restraint

Since the invasion, Russia has infrequently attacked energy infrastructure. Ukraine has also avoided attack on energy infrastructure. Their patterns of behavior demonstrate a norm of mutual restraint.[8] Thus far, mutual restraint appears to be in the interest of both sides. Of course, Russian incentives for restraint may change at any time.

Russia has attacked generating plants, but attacks with the purpose of destroying the capability to generate electricity have been few; Kremenchuk TPP, Sievierodonetsk CHP, and Okhtyrka CHP are examples of destroyed plants. Partial attacks include Zaporizhzhia NPP (attack on administrative building), Vuhlehirska TPP (attack on administrative building and fire on plant premises), Trypilska TPP (attack within plant grounds) and Kyiv CHP-6 (possibly an unintentional strike). Russia captured Kakhovka Hydro Power Plant (HPP) without damage and the plant has continued to operate. The extent of damage from attacks on Kryvyi Rih TPP is not publicly available. Russia has generally avoided attacks on high voltage transmission substations and has not attacked critical nodes of the Ukrenergo transmission system. Russia has avoided attacking high-pressure gas transmission pipelines.

Like Russia, Ukraine has demonstrated restraint. Ukraine has not destroyed generating plants when retreating. In November 2021, former foreign minister Vladimir Ohryzko warned that Ukraine could target Russian nuclear plants with missiles (Allen, 2021), but there have been no public reports of Ukraine attacking electricity infrastructure inside Russia since the invasion. Nor have there been public reports of Ukraine attacking generating plants inside Crimea or the Temporarily Occupied Territories. However, Ukraine has attacked petroleum supplies within Russia, according to the Tass news agency.

---

8   Other observers have also noticed a pattern of Russian restraint. For example, see the May 2022 New York Times article, "Russia's War Has Been Brutal, but Putin Has Shown Some Restraint. Why?" (Troianovski, 2022).

Russia's restraint is consistent with the unratified Transit Protocol of the Energy Charter Treaty. Russia advocated this Protocol to protect its commercial interests in energy transit after the breakup of the Soviet Union. Under the Protocol, Contracting Parties (signatories) would have had the obligation to ensure that energy transit is not interrupted. The Transit Protocol specifically defines energy "transport facilities." These facilities include "high-pressure gas transmission pipelines" and "high-voltage electricity transmission grids and lines." (Energy Charter Secretariat, 2002)

Keeping Infrastructure Intact

Keeping hard-to-replace infrastructure intact, especially hydroelectric and nuclear plants, is an apparent objective of Russia. The Kakhovka hydropower plant with capacity of 357 megawatts was captured intact on the first day of the invasion. Zaporizhzhia NPP with capacity of 6,000 megawatts was captured intact soon after the invasion, on March 4. Damage during the attack on the Zaporizhzhia NPP was minimal. Both plants have continued to operate and feed power into Ukraine's electric grid.

If Russia's eventual intent is to capture much of Ukraine, it has incentive to avoid damage to energy infrastructure. Reconstruction would be an arduous and extraordinarily expensive process. For example, the Dnieper system of hydroelectric facilities has taken a century to develop. Ukraine's nuclear power plants took decades to build. Damage to Ukraine's natural gas transit and storage system could take years to repair. And if Russia does not succeed in its intent to capture territory, Ukrainian infrastructure can still support transmission of Russian energy to Europe.

While Ukraine certainly opposes Russian capture of its energy infrastructure, a Russian incentive of keeping infrastructure intact might nonetheless be consistent with Ukraine's interests. In Russia's apparent mindset, if Ukraine's generating plants were to be captured intact—especially hydroelectric and nuclear power plants—their electricity could be supplied to Russia or sold to Europe. President Putin is presumably aware that Europe has an impending shortage of baseload power, because he can see coal and nuclear plants being shut down in Germany and other countries without replacement by dispatchable generation. In fact, potential sale of baseload power to Europe—at high prices during electricity shortages—may be one of the primary reasons that Russia invaded Ukraine.

Transmission of Natural Gas

Russia has an incentive to preserve Ukraine's electric grid because this infrastructure is interdependent with pipelines that transmit natural gas to Europe. As long as Russia's economy profits from the sale of natural gas, and pipeline transits outside of Ukraine are sanctioned or constrained, the incentive to preserve Ukraine's gas transmission system—and the electric grid that supports it—will remain. At current prices, Russia's Gazprom could receive approximately US$20 billion of annual revenues from gas transmitted through Ukraine's pipelines.[9] Moreover, because the price for natural gas in Europe has risen substantially over the past year—€144/MWh at the Title Transfer Facility (TTF) on June 30, 2022 vs. €22/megawatt per hour (MWh) a year earlier—the total value of gas transiting Ukraine has risen since the invasion.

To support natural gas transmission to Europe, Russian has incentives to keep Ukraine's other critical infrastructures reliably operating. The critical infrastructures most interdependent with the pipeline system are electricity and telecommunications.

Ukraine's natural gas pipeline system depends on electricity for its coordination and control. The pipeline system has hundreds of compressor and valve stations that are essential for its operation. Readings of gas pressure, flow, and temperature are transmitted to centralized control centers; in turn these readings are used to adjust compressor settings and valve positions.

Telecommunications are an integral part of the control system for natural gas pipelines, but telecommunications infrastructure relies on power from the electric grid. Without grid power for telecommunications, operation of Ukraine's pipeline system will become unreliable and hazardous. Risk of pipeline overpressure will increase. If pipeline explosions were to occur, natural gas transmission from Russia through Ukraine could halt entirely.

Prevention of Infrastructure Accidents

All countries seek to avoid infrastructure accidents. Infrastructure accidents impact human populations, cause economic losses, and harm the environment. It is in Russia's self-interest to avoid large-scale infrastructure accidents in Ukraine.

Hydroelectric dams depend on electricity and on-site staff for the safe operation of their turbines, gates, and spillways. If Ukraine's electric grid were to collapse and the dams were to be abandoned, uncontrolled water flow could result in self-destruction of the dams.

If grid power is interrupted to Ukraine's nuclear power plants and their fuel for backup generators runs out, reactor meltdowns and spent fuel pool fires could

---

9   A rough estimate of the value of gas transiting Ukraine can be determined by multiplying the flows through the Sokhranivka and Sudzha interconnector stations by the Title Transfer Facility price. Using the price and flows at the end of June 2022, this figure is about US$25 billion (€23.8) annually. In January 2022, Ukraine gas transit fees were estimated at US$1.2 billion annually (Harper, 2022). Subtracting Ukraine transit fees, Ukraine storage fees, and transit fees between Ukraine's exit point and the TTF in the Netherlands, a reasonable estimate of potential Gazprom revenues for gas transiting Ukraine would be US$20 billion annually. Much of natural gas transiting pipelines is sold under long term contracts rather than at spot market prices, such as the TTF price. Over time, contract prices tend to converge with increased spot market prices.

result. Releases of radiation could extend to Russia and to its allies as well—especially to Belarus with the Rivne nuclear plant near its border with Ukraine.

Ukraine's nuclear power plants have a design that requires power from the electric grid for operation of their cooling and safety systems. It is a common misconception that nuclear plants can operate at low power and therefore supply their own electricity when off-site power from the electric grid is unavailable. Instead, when grid power is interrupted, backup diesel generators must supply power to safety systems, including motorized pumps for reactor and spent fuel cooling. The fuel duration for backup diesel generators at Ukraine's nuclear plants is approximately seven days. Without grid power—and with backup fuel exhausted—residual heat after reactor trips could cause core meltdowns.

Water-filled pools are necessary for cooling of spent fuel rods from reactors before they are transported to nuclear waste facilities. Water in the pools shields radiation from the rods; otherwise, the rods are highly radioactive, and it is impossible for humans to be nearby. Electric power for cooling pumps is necessary to keep water in spent fuel pools from boiling. Without electric power, water in spent fuel pools would heat up and boil off. As water vaporizes, the level of water in the pool declines, eventually exposing the zirconium cladding of the fuel rods to air. The time for water to boil off would be on the order of a few weeks, assuming no makeup water is added to the pools.

If the fuel rods are recently removed from the reactor core and still hot, the junction between air, water, and zirconium provides the conditions for a strongly exothermic chemical reaction akin to a fire. Heat can vaporize the centers of the rods, releasing massive quantities of radioactive particles.

While the design of Ukrainian reactors has both core and spent fuel pool containment, violent explosions during past reactor meltdowns have breached containment vessels.[10] There can be no certainty that containment vessels would hold if Ukraine's nuclear reactors did not have off-site power from the electric grid.

Avoidance of Cross-Border Migration

Approximately two-thirds of Ukraine's population lives in urban areas. Without electricity, it is impossible for Ukraine's cities to support large populations for more than a few weeks in summer or a few days in winter. Water and sanitation systems cease functioning without grid power; pandemics from diseases such as cholera and dysentery become more likely. Food cannot be cooked, except on portable burners or wood fires. In the winter, unheated buildings become uninhabitable. Already, wartime experience in the city of Mariupol shows that when critical

---

10 The 2011 events in Fukushima, Japan demonstrated how interruption of off-site power can cause reactor explosions and breach of containment vessels. The spent fuel pools for Fukushima's reactors were not in containment vessels.

infrastructures go down, residents evacuate; for those that remain, significant casualties occur.

If the Ukrainian electric grid were to collapse over a wide area, millions of urban residents will attempt to evacuate. Thus far, evacuations of urban populations have been possible, because temperatures have been mild, transport systems have continued to operate, and outside assistance has been available. Regions of Ukraine with still functioning infrastructure have been a haven for evacuees.

During a wide-area electric grid collapse, all urban areas in Ukraine will become simultaneously uninhabitable. Residents who shelter in place will risk death; evacuation to rural areas and migration across international borders will be the only remaining options.

Russia may ultimately (and imprudently) adopt a war strategy to cause human migration to EU countries; intentional collapse of Ukraine's electric grid could be a part of this strategy. Already in Syria, Russia has demonstrated that migration can be a weapon of war. However, in the event of electric grid collapse, migration patterns would be not only flow west to EU countries, but also east to Russia and north to Belarus. Mass migration to Russia or its ally Belarus would place a resource burden on these countries—and could be politically destabilizing. Because of the potential burden of mass migration on its own society, Russia has an incentive to prevent an electric grid collapse in Ukraine. Evacuating Ukrainians may deliberately damage natural gas infrastructure, causing hardship to both Russia and Europe.

Electric Grid Scenarios

A number of scenarios are plausible for Ukraine's electric grid post-invasion. These scenarios range from the benign scenario of reliable operation to a scenario of cascading collapse followed by blackstart failure.

Nodal, relay-controlled electric grids are vulnerable to cascading collapse from unintentional failures or targeted attacks. (Cetinay, 2018) Gradual degradation of fuel supplies and resulting generation shortfalls can also cause electric grid collapse. Even before the invasion, a December 2020 analysis by a Ukrainian climate and energy policy expert warned of "cascading collapse of the power grid" in winter months due to generation shortfalls. (Savytskyi, 2020). Impaired generation and transmission capacity may prevent prompt grid restoration.

Reliable Operation

The most benign scenario for Ukraine's electric grid is its continued reliable operation, with localized outages in areas under Russian attack or occupation. Under this scenario, Russia will refrain from attacking key transmission nodes and control centers, which could cause cascading collapse. Continued supply of fuel for generating plants would allow their dispatch by Ukrenergo, the system operator. Thermal power plants could be taken offline when fuel is temporarily unavailable. When in-country generation is less than electricity demand, power could be imported from Poland or other interconnected European countries.

Reliable grid operation would support interdependent infrastructures, including telecommunications; petroleum import, storage, and distribution; natural gas transmission, storage, and distribution; coal mining and transport; and railways. Grid operation would also support infrastructure for human needs: water treatment and sanitation, food production and distribution, and residential heating.

The benign scenario of reliable operation could be possible through the summer and early fall of 2022. However, if the war persists as winter approaches, continued reliable operation of the Ukrainian electric grid will become increasingly in doubt. Continued Russian attacks will disrupt fuel supplies and place thermal power plants into forced outage. Less generation resources combined with increased demand will make it increasingly difficult for the system operator to balance generation with demand. In areas not served by CHP plants, generating plants and boiler-only heating systems will compete for scarce fuel. Cold weather will increase electricity demand, especially in regions with shutdown CHP plants.

Rolling Blackouts

Fuel shortages causing rolling blackouts are a second scenario for Ukraine's electric grid. If the war persists into the winter, fuel shortages could make rolling blackouts an increasingly likely scenario. Due to reduced supply from Russia, Ukraine will compete with EU countries for limited supplies of natural gas, coal, and fuel oil.[11] Fuel shortages for thermal power plants—mostly coal-fired and gas-fired plants—could cause intermittent shutdowns. With flexible generating plants in forced outage, the transmission system operator would be forced to maneuver large thermal plants using start/stop operation. Stress on antiquated equipment not designed for frequent cycling can lead to mechanical breakdowns. Supply chain issues could make it difficult to maintain and repair generating plants. Faced with similar fuel constraints, capacity shortfalls in Europe's ENTSO-E grid could reduce imports as a source of electricity reserves for Ukraine.

Under the fuel shortage scenario, deficits in dispatch capacity would require the transmission system operator and local distribution providers to impose rolling blackouts. Rolling blackouts may occur only during periods of high electricity demand—for example, when weather is cold, and people are using resistive

---

11 As implied by the Government of Ukraine in its production and consumption estimates, imported natural gas to meet the production gap will decline by about one-third from 2021 to 2022, from 7.5 bcm to 5.0 bcm. However, because of the sharp reduction in Russian gas transmitted through Ukraine, the "resold share" of gas flowing through the Sokhranivka and Sudzha interconnections could go up substantially—from 18% to 32%—assuming that Poland, Slovakia, and other European countries are still willing to "resell" the gas to Ukraine.

heaters. Alternatively, increases in electricity demand around daily demand peaks could require regular blackouts.

Rolling blackouts would disrupt the operation of other critical infrastructure, especially infrastructure that requires continuous electric power. If blackouts become frequent or long duration, cities heated by CHP plants may need to be evacuated.

Suffering under rolling blackouts, some inhabitants will decide to evacuate to areas with reliable power while others will attempt to operate portable generators. Portable generators are woefully inefficient; their widespread use will consume scarce petroleum supplies and thereby degrade Ukraine's warfighting capability.

Cascading Collapse

A third scenario for Ukraine's electric grid is cascading collapse. For interconnected electric grids, the physics of electricity dictate that supply must be precisely matched with demand. Under normal operation, automatic generation control (AGC), generation dispatch, and transmission flow control allow the power balance to be maintained. Abrupt disconnection of generating plants, transmission lines, or customer load causes power flows to shift. Electricity imbalances surge through the system. Protective devices at grid substations—so-called "relays"—are designed to automatically disconnect customer load, bringing supply into balance with demand. However, sometimes rapid and uncoordinated relay tripping occurs. Much like a mountain avalanche, relay trips and power surges build on themselves, resulting in complete disconnection of customer load from generation—a so-called "cascading collapse."

Even when a cascading collapse results from an accidental cause, these events can have major societal impact. For example, in August 2003, tripping of a transmission line in the Eastern Interconnection of the United States and Canada resulted in a cascading collapse affecting 55 million people; this collapse was initiated by a sagging transmission line contacting a tree branch. In September 2003, a cascading collapse in Italy and parts of Switzerland affected 57 million; this collapse was initiated by tripping of a single transmission line that flashed over towards a tree. (Sforna, 2006) In 2012, 2014, and 2015 cascading collapses impacted 620 million, 150 million, and 140 million people in India, Bangladesh, and Pakistan, respectively. Cascading collapses have simultaneous effects on infrastructure throughout large regions, greatly impacting societal functioning.

As the war persists, the probability of cascading collapse for Ukraine's electric grid increases. Worldwide experience with interconnected power systems shows that cascading collapse can occur from accidental events, such as a transmission line contacting a tree branch or misoperation of a circuit breaker. Decreases in reserve generating capacity and increased frequency of grid disruptions raise the probability of cascading collapse. Competition for fuel resources and forced outages of generating plants will reduce Ukraine's reserve capacity in coming months.

Unscheduled plant outages, substation transformer fires, and customer load sheds are all examples of grid disruptions that have occurred since the Russian invasion—and such events are likely to recur as the war persists. Russian artillery and missiles have damaged generating plants, transmission lines, and distribution substations, but fortunately none of these disruptions have resulted in cascading collapse. Coordinated attack on the transmission system is a category of grid disruption that could cause cascading collapse. Likewise, attack on key nodes of Ukraine's natural gas transmission system could cause its collapse, impacting fuel supplies for gas-fired generators. No coordinated attack on Ukraine's electricity or gas transmission systems has occurred thus far—and this appears to be a deliberate choice of Russian war strategists.

Blackstart Failure

"Blackstart" is the process of restoring an electric grid without external power. Blackstart is initiated by generating plants that can restart without off-site power and without being connected to load. Throughout the blackstart process, grid operators must precisely balance electricity generation with demand or secondary collapse will occur. During successful grid restorations, grid operators have generally completed the process in 24 hours or less. For example, the September 2003 blackout affecting all of Italy was completely restored in 18 hours and 12 minutes. (Sforna, 2006) The August 2003 Northeast Blackout in the United States and Canada was restored within 24 hours.

Although reliable blackstart is essential, operator training for its complex sequence can never be practiced in real-world conditions, because electric grids are never deliberately collapsed for practice. Blackstart initiates with flexible, fast ramping generating plants, generally hydroelectric plants and aeroderivative gas turbines. Large thermal plants—nuclear power plants and coal-fired units—are "cranked" through transmission lines. As grid restoration progresses, grid operators connect customer load in an incremental manner.

Most electric generating capacity cannot be used to initiate blackstart. The pumps, blowers, and fuel handling equipment in thermal power plants typically consume 5-10% of the plant's generating capacity. It is not economically feasible to install and maintain auxiliary generators of this capacity; therefore, large thermal plants cannot initiate blackstart. Solar and wind power cannot be dispatched to balance generation with demand and therefore are not viable blackstart resources. The long transmission lines used for imported power make it a non-optimal blackstart resource.[12]

---

12 Both real power and "reactive power" are required for grid restoration. Reactive power losses occur

The technical characteristics of Ukraine's grid as currently configured imply a possibility for blackstart failure. Aeroderivative gas turbines and combined cycle gas turbines with fast-start capabilities are rare in Ukraine, with little remaining capacity of these types after Russian capture of the Donbas region. While Ukraine has significant hydroelectric generation capacity for blackstart, these plants are often distant from the thermal generators which must be "cranked." Moreover, river flow in Ukraine is highly seasonal, with less flow in winter months—the same time that grid collapse is more likely. Approximately three-quarters of the generation capacity of Ukraine are thermal plants with steam turbine technology that requires hours of "cranking" to achieve operational temperatures. In recent years, substantial investment has been made in Ukraine's wind and solar generation, but neither of these resources can be used in the initial stages of grid restoration. Ukraine's gas-fired steam turbines are interdependent with the natural gas system—and without electricity for its coordination and control, pipeline gas could be impeded during grid restoration.

During blackstart, Ukraine's four nuclear power plants with 13.8 gigawatts of capacity—a source of electric grid strength in normal times—will become an operational challenge. Grid collapse will cause tripping and emergency shutdown of nuclear reactors. Neutron poisoning in the cores could prevent restart for two days or more.[13] "Cranking" from hydroelectric plants and restarted thermal plants will be required to blackstart the nuclear plants. The standard for emergency power at nuclear plants is seven days; once this duration is exceeded, efforts to truck in replacement fuel and otherwise prevent reactor meltdowns will consume scarce societal resources.

The phenomenon of "cold load pickup" will be a particular impediment to blackstart of Ukraine's electric grid during winter months. When grid power is interrupted during cold weather, the internal temperatures of residences and commercial buildings decline. At the point in time when power is restored, the sudden inrush of electricity into heating systems and appliances can trip grid protective devices—and again cause disconnection of customer load. (Friend, 2009) During winter blackouts in the United States, it has taken grid operators multiple attempts over several days to restore power for distribution feeders affected by cold load pickup.

Delayed blackstart can result in catastrophe, because as delays occur, grid restoration becomes progressively more challenging. Modern electric grids depend on telecommunications for their coordination and control. In turn, telecommunications equipment ultimately relies on power from the electric grid. The common

---

in the long transmission lines used to import power, reducing its utility in grid restoration.

13   For duration of neutron poisoning, the experience of nuclear plant operators in the U.S. is instructive. After the 2003 Northeast Blackout, it took tripped nuclear reactors a full week to return to 100 percent power. After tripping during the February 2021 blackout in ERCOT, it took the South Texas Nuclear Project reactor two days to restart and another 24 hours to return to full power.

standard for backup power for commercial telecommunications systems is 24-72 hours. (Popik, 2017). ENTSO-E requires 24 hours of backup power for voice communication systems used in blackstart.[14] (European Union, 2017) Substation batteries store energy for just a few cycles of circuit breaker operation. Once backup energy sources are depleted, grid restoration must be manually performed by on-site personnel using satellite phones and line-of-sight radios. Satellite phones and their communications channels will be in short supply during emergencies.

Among system operators and electric grid regulators, there is growing realization that grid restoration after system collapse may take longer than 24 hours, with attendant impacts on human populations. For example, the U.S. state of Texas experienced Winter Storm Uri in February 2021. Low temperatures caused increased demand for electricity, fuel shortages at generating plants, and mechanical breakdowns. Insufficient generation reserves required rolling blackouts and nearly resulted in complete system collapse. After the crisis, ERCOT CEO Bill Magness testified before the U.S. House Energy and Commerce Committee:

> *Avoiding a complete blackout is critical. Were it to occur, the Texas grid could be down for several days or weeks while the damage to the electrical grid was repaired and the power restored in a phased and highly controlled process...As terrible as the consequences of the controlled outages in February were, if we had not stopped the blackout, power could have been out for over 90% of Texans for weeks.*

Policy Recommendations

The Government of Ukraine and its utilities have kept the electric grid operating during wartime, a substantial accomplishment. At the time of this writing, there have been no persistent outages except those caused by attack on local transmission lines and distribution facilities. The transmission system operator, Ukrenergo, continues to perform daily dispatch. The electricity market still operates. During the April 24-25 missile attack on six railroad substations, system operators were successful in preventing cascading collapse. For the most part, generating plants have been supplied with fuel. Plant operators—even those at the occupied Zaporizhzhia NPP—have stayed on station. Line crews have risked their lives to restore power. On March 16, the Ukraine grid interconnected with the ENTSO-E grid—a year ahead of the 2023 plan—and has continued to operate reliably. The policy challenge for Ukraine is to continue this outstanding performance. Recommendations are below.

---

14 In 2017, the EU established a network code on electricity emergency and restoration that specified coordination of blackstart plans among ENTSO-E members. It is notable that the EU code specifies a minimum of 24 hours of backup power for voice communications used in grid restoration, "in case of total absence of external electrical energy supply."

Consider Russian Incentives

Ukraine should continually evaluate Russian incentives to preserve energy infrastructure. In fact, communications between Russia and Ukraine continue on a variety of levels. Past experience during the 2014-2022 Ukraine-Russia conflict in Donbas shows that infrastructure incentives can be aligned.

Aside from eliminating Ukraine as a nation and capturing its territory, a strong Russian incentive is keeping energy infrastructure intact. A system of mutual restraint has emerged. To support this incentive, Ukraine could leave energy infrastructure intact when retreating and expect the same of Russian forces. Ukraine could refrain from targeting energy infrastructure in Russian-controlled regions (and Russia itself) and expect the same from Russia. Ukraine should avoid direct military assault of the Zaporizhzhia nuclear plant currently in Russian hands.

Gas transmission through Ukraine reinforces Russian incentives for profit. Already, Ukraine is advocating for Russian shipments of natural gas to Europe to go through Ukrainian pipelines rather than the Nord Stream 1 pipeline. Reliable transmission of natural gas requires a reliable electric grid.

Prevention of infrastructure accidents should be an imperative for all countries. An electric grid collapse would dramatically increase the chance of infrastructure accidents. Ukraine should continue its demands for Russia to refrain from military actions that could cause accidents at nuclear plants under Ukrainian control. There should be an absolute prohibition on attacking hydroelectric dams.

Ukraine's actions to preserve reliable operation of its electric grid and other energy infrastructures will reduce the chance of disruptive human migrations. Foreign aid spent to reinforce Ukraine's electric grid would cost far less than money spent on assisting cross-border refugees; the Government of Ukraine should make this point to allies through public statements and diplomatic communications.

Enhance Electric Grid Resilience

Ukraine and its allies can take a number of near-term actions to enhance the resilience of its electric grid to disruption and increase the prospects for successful restoration if a collapse were to occur. Potential steps include adding reserves from electricity imports, adding flexible generation, managing consumer demand, and establishing contingency plans for quick replacement of damaged equipment.

Ukraine has already enabled electricity imports by interconnecting with Europe's ENTSO-E grid. This interconnection will provide energy during generation shortfalls and also enhance resilience to grid disruptions. Planned installation of reactive power compensation devices (STATCOM) will make imported power more useful for both normal operations and blackstart. When generating plants break down and cannot be repaired, utilities should consider using them as synchronous condensers to bolster reactive power.

Ukraine has a deficit of maneuverable, fast-response generation, especially during winter months and in regions distant from hydroelectric plants. Mobile gas-turbine generators are a common solution for stressed electric grids. These generators can provide spinning reserves to respond to grid contingencies and also initiate blackstart. Mobile gas-turbine generators have capacity of 10-50 megawatts per unit; run on natural gas, propane, and fuel oil; and can generate power in as little as two days after arrival.

Management of consumer demand will be important, especially during winter months. For example, approximately 20 percent of Ukrainians depend on district heating from CHP plants. If fuel supply is interrupted to CHP plants, or CHP generation is reduced for grid maneuverability, electricity consumers will be tempted to use plug-in resistive heaters. If 1.7 million households use plug-in heaters of 1,500 watts each during the Siberian Anticyclone, the additional load will be 2.5 gigawatts—an approximate 15% increase to winter peak demand.[15][16] A public communication campaign to reduce consumer demand during grid emergencies will be critically important.

When substation transformers are destroyed by accidental cause or deliberate attack, manufacturing and installation of replacements takes 1-2 years in normal times. Likewise, substation circuit breakers have ordering lead times of 1-2 years. Ukraine should acquire and pre-position spare transformers and circuit breakers. Ukraine can also acquire truck-deliverable, interchangeable recovery transformers, similar to the 600 mega volt amperes (MVA) RecX design developed under contract to the U.S. Department of Homeland security. (Electric Power Research Institute, 2014) Portable substations, including transformers, are also commercially available and could be pre-positioned in Ukraine.

Plan for Reliable Grid Restoration

A top priority for Ukraine should be the development of a robust grid restoration

---

[15] There are 16.8 million household consumers of electricity in Ukraine. Approximately 20% of households rely on CHP plants for district heating. The Ukraine Minister of Energy disclosed that 50% of heating capacity has been lost. This implies that approximately 1.7 million households could use resistive space heaters as a replacement for CHP heat (16.8 million household electricity consumers * 0.20 * 0.50=1.68 million). Peak demand in June 2022 was 12 gigawatts but could be 15-20 gigawatts in winter 2022-2023.

[16] Cold weather from Winter Storm Uri impacted the ERCOT system in Texas in February 2021. The unexpected increase in peak demand over the ERCOT generation adequacy plan is illustrative of the issues the Ukraine IPS could have during the Siberian Anticyclone. Starting on February 14, temperatures in Texas were much colder than normal. The ERCOT system operator planned for a winter peak load of 75.2 gigawatts, but use of resistive heating caused an additional 9.6 gigawatts of demand over the planning scenario. ERCOT imposed rolling blackouts that impacted approximately one-third of electricity consumers at their peak on February 15. (Popik, 2021)

plan. While Ukraine's fundamental electric grid configuration cannot change in the near-term, utilities can take steps to increase the probability of successful restoration. Potential steps include development of a restoration plan that is coordinated with ENTSO-E; simulator practice of blackstart by grid operators; coordination of the restoration plan with natural gas pipeline operators and telecommunications providers; acquisition of satellite phones for transmission system operators, plant operators, and distribution providers; stocking of fuel oil at dual-fuel generators; stocking of larger-than-normal reserves at coal-fired generators; installation of reactive power devices to allow imported power to be used in blackstart; installation of mobile gas-turbine generators adjacent to large thermal plants; stocking of spare high-voltage transformers and circuit breakers; maintenance of high reservoir levels at hydroelectric dams and pumped storage plants; and operation of nuclear plants at power levels that would reduce restart delays from neutron poisoning.

Ukraine has continued to operate an electricity market during the Russian invasion. Advance coordination with merchant generators would build resilience and aid grid restoration. Potential steps include ancillary services contracts for generation reserves, reactive power, and blackstart services; tests of blackstart generators under realistic conditions (without auxiliary grid power); payment to generators for stocking of fuel oil and coal reserves at generating plants; payment to gas pipeline operators for firm capacity during restoration; and Reliability Must Run (RMR) contracts with generators (payments outside of the electricity market). Contractual and operational provisions should be made for suspension of the electricity market under emergency conditions.

Electricity consumers can assist with grid restoration if expectations are publicly communicated before a blackout.[17] Potential steps include disconnecting (or unplugging) resistive space heaters; unplugging refrigerators and freezers; and generally turning off lights and appliances until after power is restored. All of these steps will reduce the challenges of "cold load pickup." A government communications program should advise home dwellers and building managers to turn off main circuit breakers before abandoning buildings.[18]

Conclusions

Ukraine has reliably operated its interconnected electric grid during wartime—a feat that is unprecedented and notably successful thus far. Ukraine's electric grid is

---

17   In the United States, some utilities have launched public communication campaigns so that their electricity consumers will mitigate the effects of cold load pickup. For an example, see "Prevent Cold Load Pickup – Unplug!" (Rio Grande Electric Cooperative, 2021)

18   In its publication "Power Outages — What to do?" the Government of Canada recommends turning off appliances, saying, "power can be restored more easily when there is not a heavy load on the electrical system" and also directs evacuating residents to "Turn off the main breaker or switch of the circuit-breaker panel or power-supply box." (Government of Canada, 2011)

increasingly stressed as the Russian invasion continues. Consideration of Russian incentives can reduce the probability of adverse infrastructure events. Military actions may degrade grid reliability or, alternatively, mutual restraint on infrastructure attacks may prevail—either outcome is possible. But as winter approaches, it is nearly certain that Europe will run short of energy and Ukraine's generating plants will compete for scarce fuel supplies. Without support from Europe and other allies, the probability of a cascading grid collapse in Ukraine increases. A cascading grid collapse followed by blackstart failure could cause a humanitarian and environmental disaster of the first order—not just for Ukraine, but also for bordering countries. Technical assistance and funding provided to Ukraine's electricity sector can significantly reduce the probability of long-term grid collapse. Moreover, the Government of Ukraine can take practical, low-cost steps on its own to increase grid resilience, such as a public communications campaign to reduce customer demand during emergencies. Support of Ukraine's electric grid during wartime has lessons for all industrialized nations dependent on critical infrastructure reliability and resilience.

Acknowledgements

The daily DiXi Group Alerts, "Russian War Against Ukraine: Energy Dimension," have been an invaluable source of information for this article. The author acknowledges the efforts of Tabor French as research analyst for this article.

Acronyms and Abbreviations

AC	Alternating Current
AGC	Automatic Generation Control
bcm	Billion Cubic Meters
CCGT	Combined Cycle Gas Turbine
CHP	Combined Heat and Power
DC	Direct Current
GDP	Gross Domestic Product
EHV	Extra High Voltage
ERCOT	Electric Reliability Council of Texas
ENTSO-E	European Network of Transmission System Operators for Electricity
EU	European Union
GTSOU	Gas Transmission System Operator of Ukraine
HPP	Hydropower Plant

HVDC	High Voltage Direct Current
IAEA	International Atomic Energy Agency
IPC	Interconnected Power System
mcm	Million Cubic Meters
MVA	Mega Volt Amperes
MW	Megawatts
MWh	Megawatt-hour
NPP	Nuclear Power Plant
PWR	Pressurized Water Reactor
SCADA	Supervisory Control and Data Acquisition
SSSCIP	State Service of Special Communication and Information Protection
STATCOM	Static Synchronous Condenser
TTF	Title Transfer Facility
TPP	Thermal Power Plant
TSO	Transmission System Operator
RBMK	High-power Channel-type Reactor
RMR	Reliability Must Run
UAH	Ukraine Hryvnia
U.S.	United States
VVER	Water-Water Energetic Reactor

Author Capsule Bio

Thomas Popik is Chairman, President, and co-founder of the Foundation for Resilient Societies. In addition to leading Resilient Societies, he serves as a principal investigator on critical infrastructures, specializing in resilience assessment, rick analysis, and economic modeling. Mr. Popik holds a Master of Business Administration from Harvard Business School and a Bachelor of Science degree in Mechanical Engineering from MIT. In his early career, Mr. Popik served as an officer in the U.S. Air Force.

References

Vladimir Afanasiev. (2022, March). "Ukraine conflict taking toll on Naftogaz operations." Upstream. Retrieved from: https://www.upstreamonline.com/product

ion/ukraine-conflict-taking-toll-on-naftogaz-operations/2-1-1188520

Felix Allen. (2021, March). "BOMB THREAT: Ukraine 'ready to strike Russian cities and nuke plants' as Putin reveals hypersonic missile will be deployed in WEEKS." The Sun. Retrieved from: https://www.thesun.co.uk/news/16896349/ukraine-ready-strike-russian-cities-nuke-plants-putin-hypersonic/

Stavros Atlamazoglou. (2022, May). "Cyberattacks quietly launched by Russia before its invasion of Ukraine may have been more damaging than intended." Business Insider India. Retrieved from: https://www.businessinsider.in/international/news/cyberattacks-quietly-launched-by-russia-before-its-invasion-of-ukraine-may-have-been-more-damaging-than-intended/articleshow/91651681.cms

Olzhas Auyezov. (2022, February). "Ukraine says Russian troops blow up gas pipeline in Kharkiv" Reuters. Retrieved from https://www.reuters.com: https://www.reuters.com/world/europe/ukraine-says-russian-troops-blow-up-gas-pipeline-kharkiv-2022-02-27/

Frank Bajak. (2022, April). "Ukraine says potent Russian hack against power grid thwarted." Associated Press. Retrieved from https://abcnews.go.com/Politics/wireStory/ukraine-potent-russian-hack-power-grid-thwarted-84034426

BBC. (2022, April). "Kramatorsk station attack: What we know so far." Retrieved from www.bbc.com: https://www.nytimes.com/2022/05/03/world/europe/russia-ukraine-war-nato.html

British Petroleum. (2022, June). "Statistical Review of World Energy." Retrieved from https://www.bp.com/content/dam/bp/business-sites/en/global/corporate/pdfs/energy-economics/statistical-review/bp-stats-review-2022-full-report.pdf

Volodymyr Boyko. (2022, April). "Imagine your city under siege with no electricity. I lived through it in Ukraine" Euromaiden Press. Retrieved from https://euromaidanpress.com/2022/04/21/imagine-your-city-under-siege-with-no-electricity-i-lived-through-it-in-ukraine/

Hannah Brown. (2022, March). "Video shows Europe's largest food warehouse on fire after 'deliberate' Russian attack" Euronews.green. Retrieved from https://www.euronews.com: https://www.euronews.com/green/2022/03/30/video-shows-europe-s-largest-food-warehouse-on-fire-after-deliberate-russian-attack

James Beardsworth. (2022, May). "Explainer: Is Russia Running Low on Missiles?" The Moscow News. Retrieved from https://www.themoscowtimes.com/2022/05/17/explainer-is-russia-running-low-on-missiles-a77704

Public Safety Canada. "Government of Canada. "Power Outages — What to do?" Government of Canada. Retrieved from: pwrtgs-wtd-eng.pdf (getprepared.gc.ca)

Hale Cetinay, Karel Devriendt and Piet Van Mieghem.(2018, August). "Nodal vulnerability to targeted attacks in power grids." Applied Network Science. Retrieved at: https://doi.org/10.1007/s41109-018-0089-9

Council of European Energy Regulators. (October 2017). "CEER Report on Power Losses." Retrieved from: https://www.ceer.eu/documents/104400/-/-/09ecee88-e877-3305-6767-e75404637087

Pin-Yu Chen and Alfred O. Hero III.(2014, November). "Assessing and Safeguarding Network Resilience to Nodal Attacks." IEEE Communications Magazine. Retrieved at: https://www.researchgate.net/publication/269098314_Assessing_and_Safeguarding_Network_Resilience_to_Nodal_Attacks

Stephanie Condon. (2022, March). "'Massive cyberattack' against Ukrainian ISP has been neutralized, Ukraine says." ZDNet. Retrieved from https://www.zdnet.com/article/massive-cyberattack-against-ukrainian-isp-has-been-neutralized-ukraine-says/

Drew FitzGerald. (2022, May). "Occupied Regions of Southern Ukraine Lose Internet Service." Wall Street Journal. Retrieved at: https://www.wsj.com/livecoverage/russia-ukraine-latest-news-2022-04-30/card/occupied-regions-of-southern-ukraine-lose-internet-service-YrGVuhNABIkQzxc099dM

DiXi Group. (2022, March). "Russian War Against Ukraine: Energy Dimension." Retrieved from: https://dixigroup.org/en/analytic/russian-war-against-ukraine-energy-dimension-daily-updating-dixi-group-alert/

DTEK. (2022, March). "Fighting for the light. How Ukrainian power grid survived 10 days of war." Retrieved from: https://dtek.com/en/media-center/news/boi-za-svitlo-yak-ukrainska-energosistema-perezhila-10-dniv-viyn/

Electric Power Research Institute. (2022, September). "Considerations for a Power Transformer Spare Strategy for the the Electric Utility Industry. U.S. Department of Homeland Security Science and Technology Directorate. Retrieved from: https://www.dhs.gov/sites/default/files/publications/RecX%20-%20Emergency%20Spare%20Transformer%20Strategy-508.pdf

Michael Elman. (September, 2000). "The 1947 Soviet Famine and the Entitlement Approach to Famines." Cambridge Journal of Economics. Retrieved from:

https://www.researchgate.net/profile/Michael-Ellman-2/publication/5208259_The_1947_Soviet_Famine_and_the_Entitlement_Approach_to_Famines/links/0deec52d5164e1f7fc000000/The-1947-Soviet-Famine-and-the-Entitlement-Approach-to-Famines.pdf?origin=publication_detail

Energy Charter Secretariat. (2022, January). "The Energy Charter Treaty, a Reader's Guide." Retrieved from: https://web.archive.org/web/20160217234855/http://www.energycharter.org/fileadmin/DocumentsMedia/Legal/ECT_Guide_en.pdf

European Union. (2017, November). "Commission Regulation (EU) 2017/2196 of 24 November 2017 establishing a network code on electricity emergency and restoration." Retrieved from: https://eur-lex.europa.eu/legal-content/EN/TXT/?uri=uriserv:OJ.L_.2017.312.01.0054.01.ENG#d1e211-54-1

Ernst & Young. (2021, August). "National Strategy to Increase Foreign Direct Investment in Ukraine." National Investment Council of Ukraine. Retrieved from: http://nicouncil.org.ua/en/strategy-eng/313-2-2-eng.html

Gas Transmission System Operator of Ukraine. (2022, May). "The actions of the occupiers led to the interruption of gas transit through the GMS Sokhranivka." Retrieved from: https://tsoua.com/en/news/the-actions-of-the-occupiers-led-to-the-interruption-of-gas-transit-through-the-gms-sokhranivka/

Gas Transmission System Operator of Ukraine. (2022, April). "The actions of Russian occupiers endanger the gas transit through GMS Sokhranivka." Retrieved from: https://tsoua.com/en/news/the-actions-of-russian-occupiers-endanger-the-gas-transit-through-gms-sokhranivka/

Global Energy Monitor. (2022, March). "Global Fossil Infrastructure Tracker." globalenergymonitor.org.

Georgii Geletukha. (2021, September). "Analysis of Ukraine's energy sector – potential for bioenergy retrofits." Bioenergy Association of Ukraine. Retrieved from: https://uabio.org/wp-content/uploads/2021/08/BIOFIT_Ukraine-22-Sep-2021_2_UABIO_Potential-for-bioenergy-retrofits-in-Ukraine.pdf

Jo Harper. (2022, January). "Can Ukraine do without Russian gas transit fees?" DW.com. Retrieved from: https://www.dw.com/en/can-ukraine-do-without-russian-gas-transit-fees/a-60552279

Interfax. "Ukrainian govt instructs Naftogaz to double gas stocks by start of heating season." (2022, June). Retrieved from: https://interfax.com/newsroom/top-stories/79987/

International Energy Agency. (2020, April.) "Ukraine energy profile." Retrieved from: https://www.iea.org/reports/ukraine-energy-profile

International Energy Agency. (2020, April.) "Ukraine energy profile." Retrieved from: https://www.iea.org/reports/ukraine-energy-profile

International Atomic Energy Agency. (2022, June.) "Update 83 – IAEA Director General Statement on Situation in Ukraine." Retrieved from: https://www.iaea.org/newscenter/pressreleases/update-83-iaea-director-general-statement-on-situation-in-ukraine

Suriya Jayanti. (2022, May). "Ukraine Is in Worse Shape than You Think." Time. Retrieved from: https://time.com/6176748/ukraine-war-economy/

Ruslan Kermach. (2022, May). "The Ukrainian electric power industry on the front line: challenges and opportunities ahead." New Eastern Europe. Retrived from https://neweasterneurope.eu/2022/05/03/the-ukrainian-electric-power-industry-on-the-front-line-challenges-and-opportunities-ahead/

Nadiya Kostyuk, Yuri M. Zhukov. (2017, November). "Invisible Digital Front: Can Cyber Attacks Shape Battlefield Events?" Journal of Conflict Resolution. Retrived from: https://doi.org/10.1177/0022002717737138

John Leyden. (2016, December). "Energy firm points to hackers after Kiev power outage." The Register. Retrieved from: https://www.theregister.com/2016/12/21/ukraine_electricity_outage/

Fred Friend. (2009, May). "Cold load pickup issues; A report to the Line Protection Subcommittee of thePower System Relay Committee of The IEEE Power Engineering Society." IEEE. Retrieved from: https://www.pes-psrc.org/kb/published/reports/Cold_Load_Pickup_Issues_Report.pdf

Saywah Mahmood. (2022, March). "Ukraine's energy security landscape mapped: where are the country's power plants located?." Power Technology. Retrieved from: https://www.power-technology.com/analysis/ukraine-power-plants/

Naftogaz. (2021, May). "Naftogaz Processing." Retrieved from: https://ugv.com.ua/en/page/pererobka

Nuclear Energy Institute. (2022). "Nuclear Power in Ukraine Fact Sheet." Retrieved from: https://www.nei.org/resources/ukraine

The Moscow Times. (2022, May). "Russia Vows to 'Relaunch' Economy of Ukraine's

Occupied Kherson." Retrived from: https://www.themoscowtimes.com/2022/05/17/russia-vows-to-relaunch-economy-of-ukraines-occupied-kherson-a77710

The Moscow Times. (2022, May). "Russia, Ukraine Trade Barbs Over Europe's Largest Nuclear Plant." Retrived from: https://www.themoscowtimes.com/2022/05/19/russia-ukraine-trade-barbs-over-europes-largest-nuclear-plant-a77725

Mattia Nelles. (2022, April). "Russia is trying to destroy Ukraine's energy sector." Spectator Australia. Retrived from: https://www.technologyreview.com/2022/03/07/1046839/how-ukraine-could-keep-the-lights-on-as-russia-attacks-its-power-supplies/

The Odesa Journal. (2022, May). "Ukraine will receive 350 thousand tons of fuel from completely new logistics routes in May." Retrieved from https://odessa-journal.com/ukraine-will-receive-350-thousand-tons-of-fuel-from-completely-new-logistics-routes-in-may/

Open Infrastructure Map. (2022, June). "All 876 power plants in Ukraine." Open Street Map. Retrieved from: https://openinframap.org/stats/area/Ukraine/plants

Morel Oprisan. (2011, May). "Prospects for coal and clean coal technologies in Ukraine." IEA Clean Coal Centre. Retrived from: https://usea.org/sites/default/files/052011_Prospects%20for%20coal%20and%20clean%20coal%20in%20Ukraine_ccc183.pdf /

Patrick Howell O'Neill. (2022, May). "Russian hackers tried to bring down Ukraine's power grid to help the invasion." MIT Technology Review. Retrieved from: https://www.technologyreview.com/2022/04/12/1049586/russian-hackers-tried-to-bring-down-ukraines-power-grid-to-help-the-invasion/

Donghui Park and Michael Walstrom. (2017, October). "Cyberattack on Critical Infrastructure: Russia and the Ukrainian Power Grid Attacks." University of Washington. Retrieved from https://icitech.org: https://icitech.org/henry-m-jackson-school-of-international-studies-university-f-washington-cyberattack-on-critical-infrastructure/

Thomas Popik. (2017, June). "Testimony of the Foundation for Resilient Societies at the June 12, 2017 Reliability Technical Conference." U.S. Federal Energy Regulatory Commission, Docket No. AD17-8-000. Retrieved from: https://elibrary.ferc.gov/eLibrary/search

Thomas Popik and Richard Humphreys. (2021). "The 2021 Texas Blackouts: Causes, Consequences, and Cures." Journal of Critical Infrastructure Policy. Retrieved from: https://www.jcip1.org/-2021-texas-blackouts.html

Anna Tsarenko. (2007). "Overview of Overview of Heating Sector in Ukraine." Center for Social and Economic Research. Retrived from: https://case-ukraine.com.ua/content/uploads/2018/09/2.pdf

Project Owl. (2022, June). "Ukraine Control Map." Retrieved from: https://www.google.com/maps/d/viewer?mid=180u1IkUjtjpdJWnIC0AxTKSiqK4G6Pez&ll=48.19156076118222%2C34.425029915307775&z=6

Reuters. (2022, May). "Ukraine to import 420,000 T of fuel in May as Russia strikes depots." Ukraine News. Retrieved from https://www.reuters.com/article/ukraine-crisis-fuel-idUKR4N2TR00R

Anna Rudyk. (2022, May). "Ukrtelecom has invested EUR 3 million in the deployment of fiber-optic networks in partnership with Slovenian." Ukraine News. Retrieved from https://ukranews.com/en/news/833171-ukrtelecom-has-invested-eur-3-million-in-the-deployment-of-fiber-optic-networks-in-partnership-with

Mario Sforna and M. Deifanti. (2006, December). "Overview of the events and causes of the 2003 Italian blackout." IEEE. Retrieved from: https://www.researchgate.net/publication/224686365_Overview_of_the_events_and_causes_of_the_2003_Italian_blackout

Oleh Savytskyi. (2020, December). "Winter is coming: will Ukrainian power sector survive it safely?" Henrich-Boll-Stiftung. Retrieved from: https://ua.boell.org/en/2020/12/09/winter-coming-will-ukrainian-energy-sector-survive-it-safely

State Nuclear Regulatory Inspectorate of Ukraine. (2022, March). "Status of Ukrainian NPP units." Retrieved from: https://snriu.gov.ua/en/timeline?&type=posts&category_id=38

Rio Grande Electric Cooperative. (2021, February). "Prevent Cold Load Pickup – Unplug!" Retrieved from: https://www.riogrande.coop/news-releases/prevent-cold-load-pickup-unplug/

James Temple. (2022, March). "How Ukraine could keep the lights on as Russia attacks its power supplies." MIT Technology Review. Retrived from: https://www.technologyreview.com/2022/03/07/1046839/how-ukraine-could-keep-the-lights-on-as-russia-attacks-its-power-supplies/

The Editor. (2021, December). "Ukraine's coal imports up 15% in Jan-Nov 2021."

The Coal Hub. Retrieved from: https://thecoalhub.com/ukraines-coal-imports-up-15-in-jan-nov-2021.html

Joao Tome and David Belson.(2022, May). "Tracking shifts in Internet connectivity in Kherson, Ukraine." CloudFlare. Retrieved at: https://blog.cloudflare.com/tracking-shifts-in-internet-connectivity-in-kherson-ukraine/

Anton Troianovski and Julian E. Barnes. (2022, May). "Russia's War Has Been Brutal, but Putin Has Shown Some Restraint. Why?" New York Times. Retrieved from nytimes.com: https://www.nytimes.com/2022/05/03/world/europe/russia-ukraine-war-nato.html

U.S. Central Intelligence Agency. (2022, May). "The World Factbook" Retrieved from: https://www.cia.gov/the-world-factbook

U.S. Energy Information Agency. (2022, January). "Ukraine Energy Profile: Important Transit Country For Supplies Of Oil And Natural Gas From Russia – Analysis." Eurasia Review. Retrieved at: https://www.eurasiareview.com/27012022-ukraine-energy-profile-important-transit-country-for-supplies-of-oil-and-natural-gas-from-russia-analysis/

State Statistics Service of Ukraine. (2022, June). "Energy supply and consumption 2020." Retrieved from: http://www.ukrstat.gov.ua/

Ukraine Ministry of Energy and Coal Mining. (2022, June). "Ukraine has submitted an offer for associate membership in the International Energy Agency." Retrieved from: http://mpe.kmu.gov.ua/minugol/control/uk/publish/article?art_id=245655306&cat_id=35109

Ukrenergo. (2019). "Report on Adequacy of Generating Capacities."

Joe Wallace and Alexander Osipovich. (2022, March). "Ukrainians Risk Their Lives to Keep Russian Gas Flowing to Europe" Wall Street Journal. Retrieved from https://www.wsj.com: https://www.wsj.com/articles/ukrainians-risk-their-lives-to-keep-russian-gas-flowing-to-europe-11646735616

Joe Wallace and Jenny Strasburg. (2022, May). "Ukraine Reduced Russian Gas Flowing to Europe Through Key Pipeline" Wall Street Journal. Retrieved from https://www.wsj.com/articles/natural-gas-prices-rise-in-europe-after-ukraine-cuts-flows-11652255011

Rob Watts. (2022, March). "Dramatic video: Fire ravages Ukraine gas facility after 'targeted' Russian attack" Upstream. Retrieved from https://www.upstreamonline.

com: https://www.upstreamonline.com/production/dramatic-video-fire-ravages-ukraine-gas-facility-after-targeted-russian-attack/2-1-1184186

Hartmut Winkler.(2022, May). "Russia's nuclear power exports: will they stand the strain of the war in Ukraine?" The Conversation. Retrieved at: https://theconversation.com/russias-nuclear-power-exports-will-they-stand-the-strain-of-the-war-in-ukraine-178250

Wikipedia. (2022, May). "HVDC Volgograd–Donbass." Retrieved at: https://en.wikipedia.org/wiki/HVDC_Volgograd%E2%80%93Donbass

Wikipedia. (2022, May). "Internet in Ukraine." Retrieved at: https://en.wikipedia.org/wiki/Internet_in_Ukraine

Wikipedia. (2022, May). "Telecommunications in Ukraine." Retrieved at: https://en.wikipedia.org/wiki/Telecommunications_in_Ukraine

World Nuclear Association. (2022, May). "Ukraine: Russia-Ukraine War and Nuclear Energy." Retrieved from: https://world-nuclear.org/ukraine-information/ukraine-russia-war-and-nuclear-energy.aspx

World Nuclear Association. (2022, March). "Nuclear Power in Ukraine." Retrieved from: https://www.world-nuclear.org/information-library/country-profiles/countries-t-z/ukraine.aspx

World Population Review. (2022, July). World War II Casualties by Country 2022. Retrieved from: https://worldpopulationreview.com/country-rankings/world-war-two-casualties-by-country

Kim Zetter. (2015, March). "Inside the Cunning, Unprecedented Hack of Ukraine's Power Grid." Wired. Retrived from https://www.wired.com: https://www.wired.com/2016/03/inside-cunning-unprecedented-hack-ukraines-power-grid/

Defense Energy Resilience and the Role of State Public Utility Commissions

William McCurry,[1,2] Lynn Costantini,[3] Wilson Rickerson,[4] Jonathon Monken,[5] Erin Brousseau[6]

[1] Senior Program Officer, Center for Partnerships and Innovation, National Association of Regulatory Utility Commissioners (NARUC)

[2] Corresponding Author, wmccurry@naruc.org

[3] Deputy Director, Center for Partnerships and Innovation, National Association of Regulatory Utility Commissioners (NARUC)

[4] Principal, Converge Strategies, LLC

[5] Principal, Converge Strategies, LLC

[6] Senior Associate, Converge Strategies, LLC

Abstract

Physical hazards such as extreme weather, and increasingly, sophisticated cyber threats, jeopardize safe, reliable operation of the power grid in the United States. More than 98 percent of military installations in the nation depend on the civilian power grid to execute military and national security missions around the world. For these installations, power outages can have devastating consequences. Public Utility Commissions are in a unique position to help bring stakeholders together and work to enhance resiliency of this defense critical energy infrastructure.

This article examines issues related to defense energy resilience and explores opportunities for PUCs to develop relationships with the Department of Defense to encourage projects that will enhance national security and provide resilience benefits to communities outside the fence line. Effective state-specific examples are highlighted.

Keywords: defense energy resilience, public-private partnership, public utility commissions, regulation

Introduction

The risks to critical energy infrastructure are rising. Extreme weather events such as fires, floods, and extended hot or cold snaps jeopardize reliable operations. Rapidly evolving cyber threats demand constant attention to avert. Consequently, the potential for destructive, long-duration power outages grows.

This threat environment has spurred the development of defense energy resilience policy and frames the current Department of Defense (DoD) approach to infrastructure investments. More than 98 percent of military installations depend on the civilian power grid and interdependent civilian utilities for communications, natural gas, and water (Narayanan et al., 2020). Power outages can affect DoD's ability to execute its military and national security missions in the U.S. and around the world. To minimize risk, energy resilience has become a central tenet of DoD's energy policy, focused on maximizing efficient energy use, expanding supply for mission assurance, and enhancing energy resilience (ASD(S), 2020a.)

In 2014, Department of Defense Directive (DoDD) 4180.01 established the need to improve energy security and "enhance the power of resiliency of installations" (DoD, 2014). In 2016, Department of Defense Instruction (DoDI) 4170.11 was updated and broadened to require that DoD "take necessary steps to ensure energy resilience on military installations ... and have the capability to ensure available, reliable, and quality power to continuously accomplish DoD missions from military installations and facilities" (DoD, 2016). The FY18 National Defense Authorization Act codified and defined energy resilience for the first time in law (H.R.2810, 2017). Today, each military branch has adopted requirements for domestic military installations to operate independently of the power grid from 7 days to 2 weeks, if necessary (ASD(S), 2020a, p. 13).

To achieve its ambitious energy resilience requirements, DoD is actively working with utilities to develop on-installation energy resilience projects and is exploring partnerships "outside the fence line," that is, outside the borders of its installations. Nonetheless, DoD must significantly accelerate the pace of investment both inside and outside the fence line to meet its energy resilience goals. DoD's success will depend, in part, on productive public-private collaboration, but few roadmaps exist. The U.S. Department of Energy (DOE) is examining opportunities to build stronger ties with DoD to advance mutually supportive energy resilience goals, as evidenced in the 2020 Memorandum of Understanding (MOU) between the DOE and DoD to address energy resilience needs of military installations and associated commercial electric grid (ASD(S)-DOE/OE, 2020) facilities. The Edison Electric Institute, an association that represents all investor-owned electric companies in the U.S., is working with the Army to pilot engagement strategies (EEI, 2019). Interestingly, state public utility commissions (PUC) are beginning to do the same.

Building Energy Resilience through Partnership

DoD is served by hundreds of different utilities, of all types, and across different electricity market structures. Investor-owned utilities (IOUs) serve over 300 major military and national security installations across the country. In most states, IOUs are regulated by the state PUCs. As such, PUCs play an implicit vital connective role between utilities they regulate and the defense communities those utilities serve.

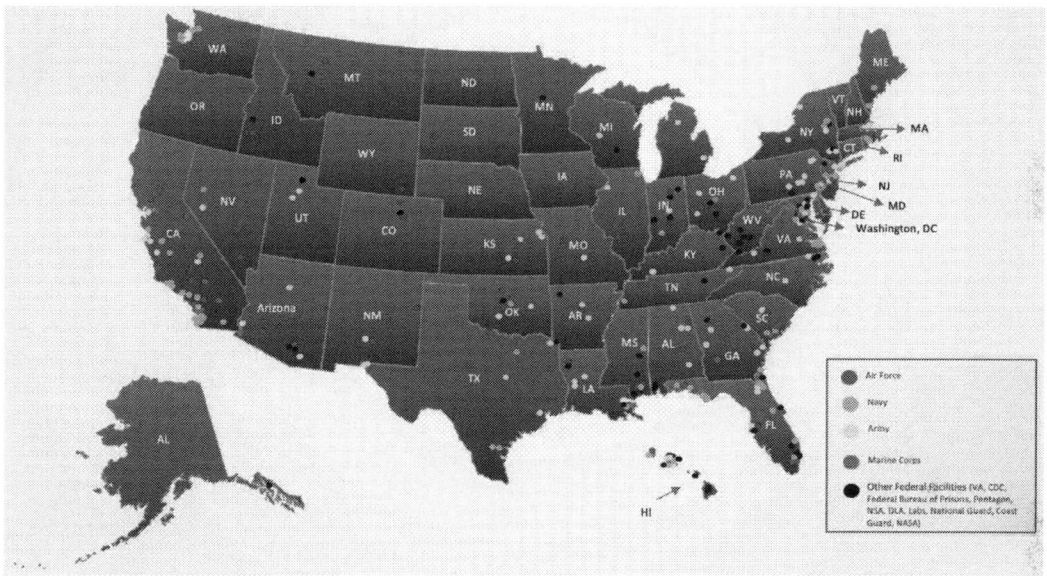

Figure 1. Department of Defense and Federal Installations Served by PUC-regulated Utilities. Source: Edison Electric Institute (2019)

PUCs are also the nexus of approval for regulated utility investments in defense energy resilience projects. Oftentimes, such projects are expected to provide resilience benefits to utility ratepayers as well. However, consideration of ratepayer investments in defense energy resilience improvements outside the fence line will require new modes of collaboration between PUCs and DoD and its federal installations that have yet to be established.

To date, most PUCs have had limited involvement with DoD related to energy resilience investments. Yet, opportunities exist for proactive outreach to discuss energy resilience priorities and the role state utility regulators might play in investment decisions around enhancing DoD energy resilience. The two notable examples that follow provide insights on the potential benefits of such collaborative engagement.

Pacific Missile Range Facility Barking Sands (Kaua'i, Hawaii)

The Hawaii Public Utilities Commission (HI PUC) approved a utility-scale solar photovoltaic (PV) and battery project (Docket No. 2017-0443) sited on land leased by a cooperative utility from the DoD (see Figure 2). The project provides firm renewable electricity to the power grid during normal operations and will serve the installation as an islandable microgrid during power interruptions and during mission critical operations. Construction on the system was completed in December 2020, and a full islanding capability test is planned in 2022.

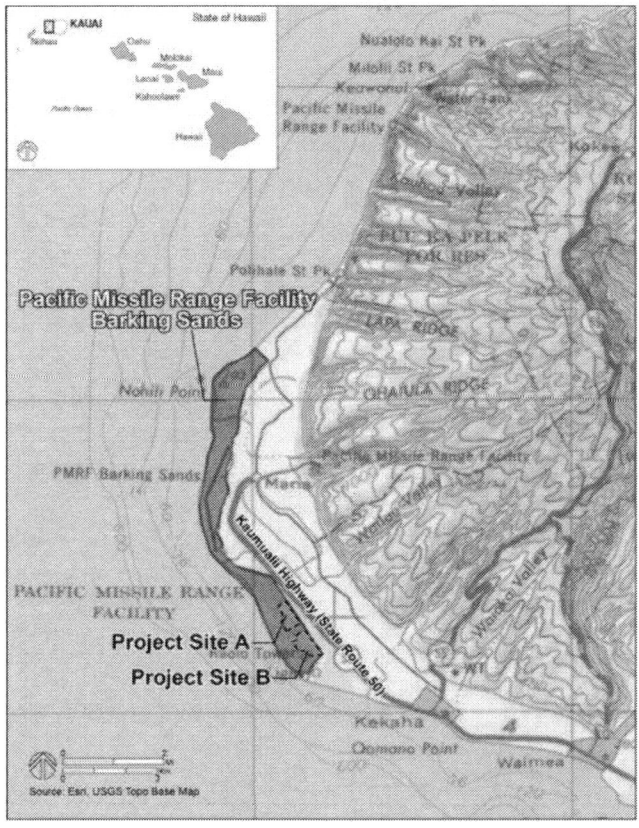

Figure 2. Sites for the KIUC Solar and Storage Facility at PMRF Barking Sands. Source: Else, 2017

In December of 2017, Kaua'I Island Utility Cooperative (KIUC) filed its initial application to HI PUC for their proposed power purchase agreement (PPA) with AES and the Pacific Missile Range Facility (PMRF) Barking Sands solar and storage project. On June 20, 2018, the HI PUC approved KIUC's application for the PPA. HI PUC concluded that the PPA was reasonable and in the best interest of the public to both add capacity and add new renewable energy generation. The price of $108.50 per MWh was found to be reasonable. HI PUC also approved KIUC's sublease to AES, KIUC's request to build an above-ground transmission line, as well as KIUC's request to spend approximately $8.87 million on the PMRF Barking Sands Substation.

Davis-Monthan Air Force Base (Tucson, Arizona)

The Arizona Corporation Commission (ACC) approved an environmental compatibility certificate for a planned project by Tucson Electric Power (TEP) to expand and upgrade the transmission system in the region surrounding Davis-Monthan Air Force Base (DMAFB) (Docket No. L-00000C-20-0007-00186). The project was undertaken to enhance service reliability for current and new customers, and in response to DoD energy resilience policies and requirements. The project will reduce the risk of transmission system overloading within the region and enhance service reliability for current and new customers. The upgrades will also accommodate the expansion of renewable energy resources in the future (TEP, 2020a). The new substation will be fed by two separate 138 kV transmission lines. The addition of a second point of entry for electrical power to DMAFB eliminates the risk to the base of a single point of failure.

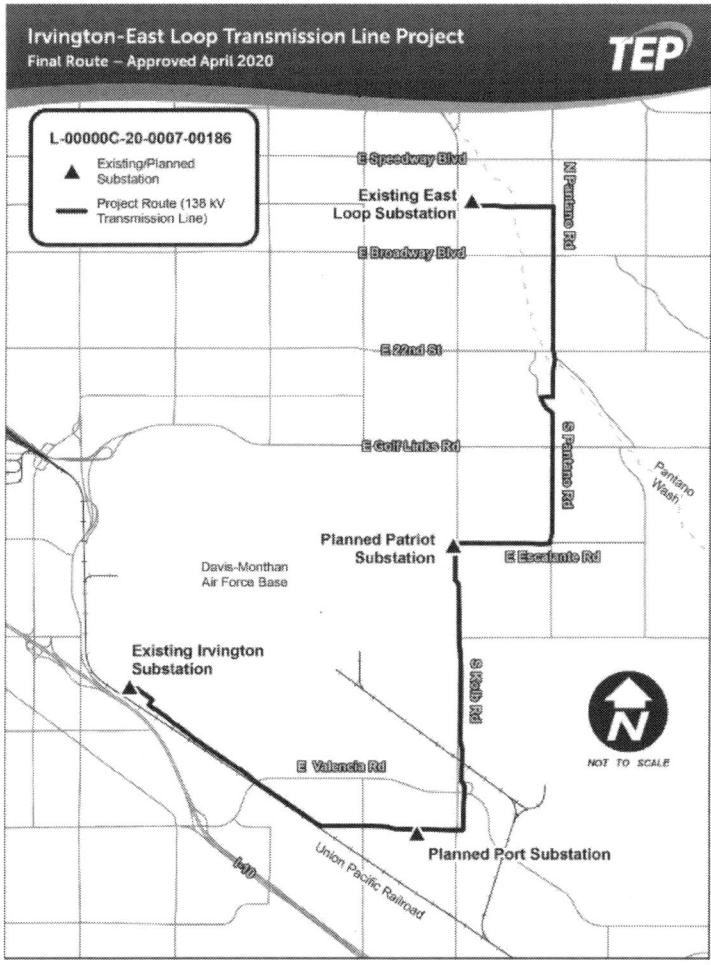

Figure 3. Irvington to East Loop 138 kV Transmission Project
Source: TEP, 2020b

On January 15, 2020, TEP applied for a Certificate of Environmental Compatibility with the Arizona Power Plant and Transmission Line Siting Committee. TEP provided testimony that "the decision to move forward with the project was driven by the needs of DMAFB to meet a new DOD mandate regarding energy resiliency" (ACC, 2020a, p. 13). In addition to emphasizing DoD energy policy, TEP also specifically cited the requirements of the Air Force Energy Flight Plan (U.S. Air Force, 2017) to increase the use of energy resiliency technologies and partnerships for critical infrastructure, eliminate single points of failure for facility energy, and eliminate energy shortfalls to improve contingency operations (ACC, 2020a, p. 83). In April 2020, the ACC issued an order approving the certificate, allowing the project to move forward. ACC found that the project was in the public interest because it would enhance the utility's ability to respond to future load growth, provide support to existing distribution substations, and "assist [DMAFB] in fulfilling the DoD … directive for enhancing energy resiliency" (ACC, 2020b).

As these examples suggest, PUCs across the U.S. may benefit from a greater understanding of the broader DoD energy resilience policy landscape as well as their specific role in fostering relationships that enhance the protection of critical infrastructure and enhanced national security. The National Association of Regulatory Utility Commissioners (NARUC) recently released a white paper describing the military's defense energy resilience policies and suggestions for opportunities for PUC engagement (Rickerson, et. al, 2021). The paper describes key steps and raises important considerations for PUCs as they begin to grapple with defense-related topics:

Assess in-state defense critical infrastructure (DCI)

Regulators can identify the military installations in their state and review the extent to which DoD bases are served by regulated utilities. This assessment could be limited to electricity infrastructure or could also include a review of how additional sectors that commissions regulate serve in-state defense facilities.

Engage with in-state DoD representatives and activity

Regulators can engage the staff of in-state military installations to identify planned or ongoing energy resilience projects that may rise to the level of commission consideration.

Initiate an investigative docket

Commissions in some states can initiate dockets that are informational in nature. Commissions could open proceedings to investigate, for example, the current status of utility and military partnership within the state, the nature and duration of outages experienced by military facilities, opportunities to include projects with military co-benefits in integrated resource planning, or options for low-cost or no-cost resilience improvements. Adding DoD sites to black start crank paths or to the bottom of load shed lists, for example, represent lost-cost ways to improve the resilience of defense facilities.

Explore the development of joint energy resilience metrics

There are multiple ongoing efforts to identify metrics that can be used for resilience planning, and for utility energy resilience (Anderson et al., 2017; Kallay, Napoleon, Havumaki et al., 2021). Regulators can review and engage with these efforts as they develop.

Convene and/or participate in value of resilience investigations

There are multiple, ongoing efforts to explore the value of resilience that are being led by the national laboratories and by the utility industry. In the near-term, regulators can engage with these initiatives to assess emerging methodologies and provide feedback on potential use cases. Regulators can also commission their own studies of the value of resilience and work with utilities to integrate them in benefit-cost analyses.

Investigate defense energy-related cybersecurity investments

Given the ambiguities around the comparative benefits and costs of cybersecurity countermeasures, regulators could work with in-state military installations to anchor investigations into the effectiveness of cybersecurity strategies.

Engage in planned or ongoing defense-relevant collaborations and explore secure communications frameworks

Pilot projects represent opportunities for commissions to engage with and learn alongside the communities, utilities, state agencies, and defense agencies. State public utility commissions can work in close conjunction with their sister agencies – state energy offices – to coordinate funding for military energy resilience projects. The California Energy Commission, for example, signed an MOU with the Department of the Navy to support renewable energy development, energy security, and energy reliability (California Energy Commission, 2016), following recommendations from the California Governor's Military Council (2015). Commissions could also provide guidance or checklists to the proponents of ongoing efforts about how to best prepare to have productive regulatory engagement as their projects mature. Future work in the defense energy resilience space will need to include policy development on how regulators can best engage in secure communications and decision making as they address the issues outlined in this article.

    Ultimately, a PUC may initiate proceedings to explore the current status of utility-military partnership within the state, the nature and duration of outages experienced by military facilities, options for low-cost or no-cost resilience improvements, and opportunities to integrate projects with co-benefits between defense customers of utility services and the surrounding community into integrated system planning.

Emerging Issues for Regulators

The field and practice of defense energy resilience remains nascent as does the role of state utility commissions within it. Going forward, regulators may proactively engage with issues related to defense energy resilience, or they may increasingly see issues related to defense energy resilience integrated into their normal course of business. In either case, there are several uncertainties and unresolved issues that regulators will need to navigate when considering defense-related topics.

Rate Recovery—Who Pays?

Although many stakeholders are active in areas related to defense energy resilience, clear responsibilities for coordinating and resourcing investments have not been established. The rapidly evolving policy landscape at the federal and state levels has created different avenues to support defense energy resilience, but each has benefits and drawbacks.

DoD funding

DoD has a massive energy budget to acquire the fuel and systems it needs to sustain global operations. DoD's budget for domestic on-base energy improvements, however, is limited compared to the need (Niemeyer, 2018). DoD relies primarily on third-party financing to acquire energy efficiency, renewable energy, and energy resilience. DoD has also not historically had the funds or the authority to make "outside the fence line" investments in community and private sector energy infrastructure on which its installations depend.

Other federal funding.

There are a broad range of funding programs from federal agencies beyond DoD that could potentially be used to support energy infrastructure projects. Many of these are described in a NARUC publication titled "Federal Funding Opportunities for Pre- and Post-Disaster Resilience Guidebook" (Monken, 2021). Funding opportunities included in the Infrastructure Investment and Jobs Act (The White House EOP, 2022), and programs such as FEMA's Building Resilient Infrastructure and Communities have budgets that are an order of magnitude larger than DoD's community programs. Defense energy resilience-related investments could be possible, but the programs serve a broad range of competing uses beyond defense infrastructure and may be difficult to secure. The Defense Access Roads Program, for example, is a joint program between DoD and the Federal Highway Administration for DoD to pay its share of the cost of public highway improvements required by defense activity (Federal Highway Administration, 2021). Similar programs for DoD to contribute to energy infrastructure do not yet exist.

Ratepayer funding

National security is a public good, and investments in electrical infrastructure to secure critical DoD bases is broadly in the public interest. Utilities are best positioned to make investments in their systems that serve DoD bases, but the extent to which in-state utility ratepayers should carry the cost of defense-relevant infrastructure is unclear. Commissions have rejected ratepayer recovery for some non-DoD energy resilience projects because they served too narrow a geographic area and did not create sufficiently widespread ratepayer benefits (Rickerson, Gillis, and Bulkeley, 2019). Similar arguments could be made to limit investments that benefit DoD installations alone.

On the other hand, DoD installations have successfully demonstrated that they are able to use on-base energy resilience systems to support grid operations during severe weather events to the benefit of regional ratepayers. Military-utility energy resilience cooperation can be a two-way street. Just as there have been examples of civilian support for military energy resilience projects, there have been multiple instances during the past several years during which military bases have used their on-base generating assets to support grid stability during severe climate events. Two recent examples involved heat waves in California in 2020 and the polar vortex of 2021:

Marine Corps Air Station (MCAS) Miramar

In August 2020, California ISO issued an alert to reduce energy demand statewide in response to a record-breaking heat wave (California ISO, 2020). MCAS Miramar in San Diego has an installation-wide microgrid that incorporates battery storage, and landfill gas, solar PV, natural gas, and diesel generation (Booth et al., 2020). During the event, MCAS Miramar activated the microgrid to reduce demand on the commercial grid. The base was able to remove 6 MW of its demand from the grid, which helped keep an estimated 3,000 homes online (Carlisle, 2020; Dockery, 2021).

Offutt Air Force Base

A polar vortex in February 2021 brought extreme cold temperatures across the southwestern and midwestern United States. The cold caused widespread power outages across the United States and Mexico and precipitated a power crisis in Texas. In Nebraska, parts of the state experienced temperatures nearing -20 degrees. In Omaha, NE, the Omaha Public Power District contacted Offutt Air Force

Base to help alleviate the strain on the grid (Starr and Kaufman, 2021). Offutt Air Force Base activated its on-base power plants and emergency back-up power systems and reduced their demand on the grid by 6 MW. The base utilized its on-site generation assets to support the grid for 75 hours (U.S. Air Force, Offutt Air Force Base, 2021).

How to Weigh Benefits: Quantifying the Value of Defense Energy Resilience

Utility investments in defense electric infrastructure can create a resilience value that accrues to a broad range of stakeholders above and beyond the economic development benefits discussed in the previous section. State commissions have emphasized the need for a quantitative resilience value to support rulemaking, rate making, and emergency planning (California PUC, 2020; MPSC, 2019). The value of resilience is typically acknowledged to be significant, but notoriously difficult to quantify. There have been many attempts to identify a resilience value that can be used to support energy decision making, but the energy industry has not adopted a standard approach (Electric Power Research Institute, 2021).

The practice of integrating energy resilience into regulatory benefit-cost analysis remains at an early stage (Kallay, Letendre et al., 2021), and NARUC research found that a quantified value of resilience has not been considered in commission energy resilience proceedings (Rickerson, Zitelman et al., 2022). Research into the value of resilience is ongoing, however, and both commissions and utility stakeholders will likely continue to attempt to integrate resilience into regulatory proceedings in the future.

Utility cost recovery may be most feasible when there are clear benefits for ratepayers above and beyond supporting national security. In the case of PMRF Barking Sands, the project will provide additional energy and capacity to the grid while helping the state cost-effectively achieve its renewable energy target. In reviewing TEP's proposal in the region around DMAFB, ACC recognized the project's broader benefits in its 2021 Order.

Cybersecurity may also have bearing on commissions' consideration of defense energy resilience. Most of the case studies and examples in this report focus on investments in physical electricity infrastructure (e.g., transmission, distribution, and generation). Commissions are increasingly being asked to consider whether the costs of utility investment in cybersecurity measures are just and reasonable, and there are many cases in which commissions have approved ratepayer investments in cybersecurity. Although asset-level investments at military installations may face scrutiny as to whether they benefit ratepayers broadly, cybersecurity investments in central automated systems that serve DoD installations would likely also create benefits for civilian customers across utility service territories.

Military Energy Resilience and Economic Development

The role of economic development within regulatory proceedings is mixed. Although regulators in many states are statutorily able to consider economic impacts as in rulemaking, some commissions are not and may invite litigation if they do so (Zitelman and McAdams 2021). Even if commissions are allowed to consider economic impacts, it does not mean that they will. Recent studies of regulatory decisions related to damage from extreme weather events, for example, found that commissions have not used regional economic impacts in their decision making (Sanstad et al., 2020).

There have not yet been cases in which regulators have specifically taken the economic impact of avoiding base closure into account when considering military energy resilience investments. However, there are some instances in which state regulators have considered base retention when evaluating military renewable energy investments. The Alabama Public Service Commission (Alabama PSC) approved a petition from Alabama Power to build and own a 10.6 MW_{AC} solar PV plant at both Fort Rucker and at Anniston Army Depot in Docket No. 32382 in 2015.

Alabama Power justified its request to partner with the Army in part by citing federal law requiring agencies to procure renewable energy and stating that the PV projects "should help [the installations] avoid unwarranted scrutiny by federal leaders (Alabama PSC, 2015a)." In its Order approving the two Army projects, Alabama PSC cited staff analysis that considered the "direct benefits associated with retaining the military bases load by supporting them in meeting federal mandates associated with renewable energy standards and the indirect benefits associated with retaining residential and commercial loads that are highly dependent on the economic impact of each military base" (Alabama PSC, 2015b). Given the energy resilience policies set by DoD within the last 5 years, there may be instances in the future in which commissions are asked to consider military energy resilience from an economic development perspective.

Next Steps

State public utility commissions play a critical role in approving a variety of utility investments for projects directly impacting defense community customers. PUCs can act as a convening entity to encourage collaborative engagement among utilities, defense customers, public agencies, and local communities. Ample opportunity exists for PUCs to develop relationships with DoD to encourage projects that will enhance national security and provide resilience benefits to communities outside the fence line.

PUCs are strongly encouraged to familiarize themselves with the strategic priorities around defense energy resilience and develop their own guidelines for engagement on these issues, particularly as these types of investments become more common. NARUC is in the process of developing several key resources to help PUCs navigate the complexities of the DoD energy policy landscape and to provide some guidance on how a PUC might engage on defense energy resilience topics. Those resources will be forthcoming later in the year.

References

Alabama Public Service Commission. (2015a). Order (Docket No. 32382). Montgomery, AL.

———. (2015b). Order (Docket No. 32382). Montgomery, AL.

Anderson, D., A. Eberle, T. Edmunds, J. Eto, S. Folga, S. Hadley, G. Heath, et al. (2017). "Grid Modernization: Metrics Analysis (GMLC1.1)" (PNNL-26541). Richland, WA: Pacific Northwest National Laboratory.

Arizona Corporation Commission (ACC). (2020a). "Notice of Filing of Direct Testimony and Exhibits of Tucson Power Company." Phoenix, AZ.

———. (2020b). Order (Decision No. 77601). Phoenix, AZ.

Booth, S., J. Reilly, R. Butt, M. Wasco, and R. Monohan. (2020). "Microgrids for Energy Resilience: A Guide to Conceptual Design and Lessons from Defense Projects" (NREL/TP-7A40-72586). Golden, CO: National Renewable Energy Laboratory.

California Energy Commission. (2016, October 13). "Navy and Energy Commission Formalize Energy Partnership." Sacramento, CA.

California Governor's Military Council. (2015). "Maintaining and Expanding California's National Security Mission." Sacramento, CA.

California Independent System Operator. (2020). "Flex Alert Issued for Next Four Days, Calling for Statewide Conservation." Folsom, CA.

Edison Electric Institute (EEI). (2019). "Department of Defense & Federal Government." Washington, DC.

Electric Power Research Institute. (2021). "Value of Resilience White Paper." Palo Alto, CA.

Else, J. (2017, February 28). "PMRF Solar Project Could Make Island Self-Sufficient." *The Garden Island*.

Hawaii Public Utilities Commission. (HI PUC). (2015). Decision and Order No. 33178. Honolulu, HI.

———. (2017). Docket No. 2017-0443. Honolulu, HI.

———. (2018). Decision and Order No. 35538. Honolulu, HI.

H.R.2810—115th Congress (2017–2018): National Defense Authorization Act for Fiscal Year 2018. (2017, December 12).

Kallay, J., S. Letendre, T. Woolf, B. Havumaki, K. Shelley, A. Hopkins, R. Broderick, R. Jeffers, and B. M. Garcia. (2021). "Application of a Standard Approach to Benefit-Cost Analysis for Electric Grid Resilience Investments" (SAND2021-5627). Albuquerque, NM and Livermore, CA: Sandia National Laboratories.

Kallay, J., A. Napoleon, B. Havumaki, J. Hall, C. Odom, A. Hopkins, M. Whited, et al. (2021). "Performance Metrics To Evaluate Utility Resilience Investments" (SAND2021-5919). Albuquerque, NM: Sandia National Laboratories.

Kallay, J., A. Napoleon, J. Hall, B. Havumaki, A. Hopkins, M. Whited, T. Woolf, et al. (2021). "Regulatory Mechanisms to Enable Investments in Electric Utility Resilience" (SAND2021-6781). Albuquerque, NM: Sandia National Laboratories.

Kaua'i Island Utility Cooperative (KIUC). (2019). "Strategic Plan Update." Kaua'i, HI.

———. (2021). "Energy Information." Kaua'i, HI.

Michigan Public Service Commission (MPSC).(2019). "Statewide Energy Assessment Final Report." Lansing, MI.

———. (2020). "MPSC Statewide Energy Assessment." Lansing, MI.

Monken, J., S. Cohen, E. Brousseau, J. Graul, et al. (2021). "Federal Funding Opportunities for Pre- and Post-Disaster Resilience Guidebook". Converge Strategies LLC on behalf of National Association of Regulatory Utility Commissioners (NARUC). Washington, DC.

Narayanan, A., J. Welburn, M. Miller, S. Li, and A. Clark-Ginsberg. (2020). "Deterring Attacks Against the Power Grid." Santa Monica, CA: Rand Corporation.

Niemeyer, L. (2018). "Statement of Honorable Lucian Niemeyer, Assistant Secretary of Defense (Energy, Installations and Environment)." Washington, DC: House

Committee on Armed Services Subcommittee on Readiness.

Rickerson, W., W. Gillis, and M. Bulkeley. (2019). "The Value Of Resilience For Distributed Energy Resources: An Overview Of Current Analytical Practices." Prepared for National Association of Regulatory Utility Commissioners. Washington DC: Converge Strategies.

Rickerson, Wilson, K. Zitelman, and K. Jones. (2022), "Valuing Resilience for Microgrids: Challenges, Innovative Approaches, and State Needs", Report for National Association of State Energy Officials (NASEO) and the National Association of Regulatory Utility Commissioners (NARUC) Microgrids State Working Group.

Rickerson, Wilson, E. Brousseau, J. Monken, M. Pringle, J. Graul, T. Calvert-Rosenberger, J. Barker (2021) "Regulatory Considerations for Utility Investment in Defense Energy Resilience" Prepared for National Association of Regulatory Utility Commissioners. Washington DC: Converge Strategies.

Sanstad, A., Q. Zhu, B. Leibowicz, P. Larsen, and J. Eto. (2020, November). "Case Studies of the Economics Impacts of Power Interruptions and Damage to Electricity System Infrastructure from Extreme Events." Berkeley, CA: Lawrence Berkeley National Laboratory.

Tucson Electric Power (TEP). (2020a). "2020 Integrated Resource Plan." Tucson, AZ.

———. (2020b). "Irvington-East-Loop Transmission Line Project." Tucson, AZ.

U.S. Air Force. (2017). "Energy Flight Plan 2017–2036." Washington, DC.

U.S. Air Force, Offutt Air Force Base. (2021). "Offutt Reduces Power Grid Load to Help Community During Frigid Temps." Omaha, NE.

U.S. Army, Office of Energy Initiatives (OEI). (2019a). "Anniston Army Depot, Alabama Solar Energy Project Provides On-Site Generation for Potential Microgrid & Supply Diversity." Washington, DC.

———. (2019b). "Fort Rucker, Alabama Solar Energy Project Provides On-Site Generation for Potential Microgrid & Supply Diversity." Washington, DC.

U.S. Department of Defense (DoD). (2014) "Department of Defense Directive 4180.01 DoD Energy Policy" Washington, DC.

———. (2016) "Department of Defense Instruction 4170.11 Installation Energy

Management". Washington, DC.

U.S. Department of Defense, Office of the Assistant Secretary of Defense for Sustainment (ASD(S)). (2016). "Installation Energy Plans." Washington, DC.

———. (2020a). "Annual Energy Management and Resilience Report (AEMRR) Fiscal Year 2019." Washington, DC.

———. (2020b). "Energy Resilience and Conservation Investment Program (ERCIP)." Washington, DC.

———. (2020c). "Fiscal Year 2021 Operational Energy Budget Certification Report." Washington, DC.

U.S. Department of Defense, Office of the Assistant Secretary of Defense for Sustainment (ASD(S)), and U.S. Department of Energy, Assistant Secretary for the Office of Electricity (DOE/OE). (2020). "Memorandum of Understanding between DoD and DOE." Washington, DC.

U.S. Federal Highway Administration. (2021). "Defense Access Road Program (DAR)." Washington, DC.

The White House – Executive Office of the President. (2022) "Building A Better America: A Guidebook to the Bipartisan Infrastructure Law for State, Local, Tribal, and Territorial Governments, and Other Parties" Washington, DC.

Zitelman, K., and J. McAdams. (2021). "The Role of State Regulators in a Just and Reasonable Energy Transition: Examining Regulatory Approaches to the Economic Impacts of Coal Retirements." Washington, DC: National Association of Regulatory Utility Commissioners.

Expanding Oregon's Vision for a Once-in-a-Generation Infrastructure Investment

Robert Parker[1] and John Tapogna[2]

[1] Director of Strategy and Technical Solutions, Institute for Policy Research & Engagement, University of Oregon, rgp@uoregon.edu

[2] Senior Policy Advisor, ECONorthwest, tapogna@econw.com

Abstract

Oregon will receive more than $5 billion in Infrastructure Investment and Jobs Act (IIJA) which will finance projects in more than 100 programs across state agencies, local governments, and tribes. In this paper, we argue that, rather than a job-creating road, water, and broadband bill, Oregon stakeholders should actively marshal IIJA funding as a critical building block for three of the State's most pressing economic issues: equity, housing, and industrial revitalization.

The Opportunity

Congress passed the $864 billion Infrastructure Investment and Jobs Act (IIJA) in November 2021. The five-year initiative represents an overdue federal investment in regional economies in Oregon and across the nation. The IIJA's aggressive implementation timetable provides advantages to states with a clear vision—and substantial prior planning—centered on the relationship between infrastructure and economic development.

The opportunity is at once simple and complex. Although the Act includes dozens of competitive opportunities and could create more than 100 new federal programs, about three-quarters of the resources, $660 billion, will pass directly to state and local governments through formula funding. And while IIJA covers a broad range of infrastructure areas, about two-thirds of resources, $591 billion, are devoted to transportation—highway, roadway, bridge, and transit programs.

The White House reports that Oregon will receive more than $5 billion in IIJA funds which will finance projects in more than 100 programs across state agencies, local governments, and tribes.[1] It's a significant spending boost: Oregon

---

1 https://www.oregon.gov/odot/IF/Pages/Governors-Infrastructure-Cabinet.aspx

state and local governments made $1.6 billion in capital outlays for highways and utilities in 2019.[2]

Table 1. Infrastructure Investment and Jobs Act (IIJA) Allocations to Oregon

IIJA formula allocations to Oregon (in billions)	
Modernize roads and highways	3.4
Upgrade public transportation	0.7
Improve water infrastructure	0.5
Connect Oregonians to the internet	0.1
Bridge replacements/repairs	0.3
Other (airports, resiliency)	0.3
Total	5.3

The Oregon Department of Transportation (ODOT) will be a key implementing agency and anticipates spending at least $1.2 billion on a variety road, bridge, transit, footpath, bike path and climate migration projects.[3] Two-thirds of the known ODOT funds are earmarked for specific projects; the remaining one-third is allocated for flexible funding. The funds will affect dozens of other programs managed by several key state agencies including Business Oregon, Oregon Housing and Community Services, Oregon Department of Environmental Quality, Oregon Health Authority, Oregon Water Resources Department, Oregon Department of Energy, Oregon Fish and Wildlife, and Oregon Department of Aviation.

In addition to the formula allocations, Oregon's state and local governments can compete nationally for resources that will be steered to economically significant bridge investments and projects that deliver substantial economic benefits to local communities. Portland's long-planned replacement of the Oregon-Washington Interstate 5 Bridge and the Port of Coos Bay's Pacific Coast Intermodal Port projects are strong candidates for competitive funding.

The IIJA: Unlocking Economic Potential in Oregon's Complex Development Environment

Infrastructure is a key economic building block, and in Oregon, a unique, statewide planning system often determines where, how, and whether it is deployed. Across Oregon and across decades, localities have navigated complex rules and secured hard-fought approvals for development. But too often, a lack of infrastructure funding has proven to be an insurmountable, final barrier. The IIJA offers a

---

[2] Source: Urban Institute Brookings Institution Tax Policy Center's State and Local Finance Data tool. Accessed on June 20, 2022 at https://state-local-finance-data.taxpolicycenter.org/pages.cfm

[3] https://www.oregon.gov/odot/Documents/IIJA-Community%20Priorities.pdf

rare opportunity to unlock economic potential and to make investments that pay returns over decades.

Infrastructure sits at the foundation of an economy. Like human, social, and natural capital, it is a core asset supporting multi-generational economic growth. A large body of research has established the long-run benefits of public infrastructure spending and concluded that most advanced countries underinvest in it.[4] Infrastructure drives economic growth by supporting land uses that facilitate productive activity. It builds community, enables social interaction, helps move goods and services to market, and provides places for people to live. The planning and deployment of infrastructure evolves over decades. Communities first identify needs in 20-year master plans. As economic opportunities come into clearer focus capital improvement plans prioritize investments in a three-to-six-year window. And then when the market is ready, a site is developed.

Through planning and the gradual addition of infrastructure, land transitions through a series of states, or transition events, that lead from greenfields to developable lots.[5] In Oregon, those transitions are governed by the state's nearly 50-year-old land use planning program. Its iconic requirement calls on every city to establish an Urban Growth Boundary (UGB) to protect resource lands outside the boundary and encourage higher density development patterns inside the boundary.

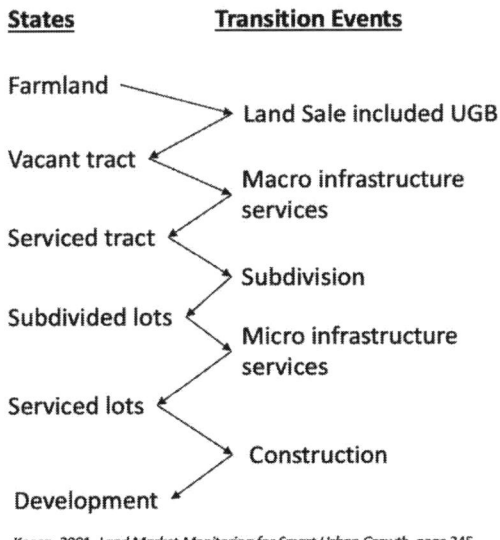

Knaap, 2001, *Land Market Monitoring for Smart Urban Growth*, page 245

Figure 1. Pathways to Development Readiness – Land States and Transition Events

---

4   See Ramey, Valerie A. (July 2020) *The Macroeconomic Consequences of Infrastructure Investment*. Working Paper 27625. National Bureau of Economic Research. Cambridge, MA. and Glaser and Poterba. (December 2020) *Economic Analysis and Infrastructure Investment*. Working Paper 21215, National Bureau of Economic Research, Cambridge, MA.

5   "Land Market Monitoring for Smart Urban Growth," Knaopp, Gerrit-Jan. Lincoln Institute of Land Policy, November 2001.

The program—through its Goal 11 on Public Facilities and Services—also requires cities and counties "to plan and develop a timely and efficient arrangement of public facilities and services to serve as a framework for urban and rural development."[6] Goal 11 and its related administrative rule establish two important requirements: (1) cities and counties must adopt public facilities plans; and (2) local governments cannot allow development of sewer systems outside urban growth boundaries or unincorporated community boundaries.

While Oregon has a well-established set of planning requirements for localities, the system largely leaves the execution and funding of infrastructure to local governments. And a lack of infrastructure funding was determined to be a key barrier in greenfield-to-developable lot transitions. An analysis of Portland-area's UGB expansions found that—14 to 18 years after the expansion— newly added areas had developed only 2.5% of their envisioned housing capacity. A subsequent, more detailed analysis of many of the same expansion areas and found that infrastructure funding and availability was a barrier to development in many of these areas.[7] Similarly, on the industrial side, a lack of infrastructure leaves prime land vacant decades after its introduction into the UGB. More than 43% of respondents rated bringing infrastructure to land within UGBs as an "extreme barrier" to housing production in Oregon; another 34% rated it as a "moderate barrier."[8]

The bottom line: Oregon's highly prescribed, statewide approach to land-use and infrastructure planning, together with the devolution of infrastructure finance, has contributed to a shortage of developable residential, commercial, and industrial lots. The IIJA does not solve Oregon's underlying land-use and financing challenges, but it does open a door to address a sizeable backlog of long-identified development opportunities in small and large communities across the State.

Ready to Act: Oregon's Early Steps on IIJA

Oregon is known for its vision and innovation in the transportation and infrastructure sectors. The 1970s conversion of the Mt. Hood Freeway into a light rail project is part of State lore and ushered in an era of investments that elevated the importance of community vibrancy and social capital in physical capital planning. Oregon was among the best stewards of federal highway dollars in past generations[9] and, with modern finance innovations like the vehicle miles tax and congestion pricing in the works, appears poised to maintain its leadership in the next. At the federal level, Oregon and the nation have been well served by Representatives

---

6   https://www.oregon.gov/lcd/OP/Documents/goal11.pdf

7   https://www.oregonmetro.gov/sites/default/files/2016/04/26/UGB%20Report%20for%20Metro%20FINAL%20-%20combined%2004%2026%202016.pdf

8   Barriers to Housing Production in Oregon, Rebecca Lewis and Robert Parker, research in progress.

9   Brooks, Leah and Zachary Liscow (October 2020). *Can America Reduce Highway Spending? Evidence from the States.* Working Paper

Expanding Oregon's Vision for a Once-in-a-Generation Infrastructure Investment

Peter DeFazio, retiring chair of the House Committee on Transportation and Infrastructure and Earl Blumenauer, a national thought leader on transit-oriented development.

The State's leadership in the area positions it well for this opportunity. Soon after IIJA's passage, Governor Kate Brown convened an Infrastructure Cabinet to coordinate with the federal government, state agencies, and local governments. Racial equity is a priority as are meeting key federal expectations of "resilience, equity, not leaving anyone behind, labor, made in America, and climate resilience."[10]

The IIJA's implementation coincides with ODOT's recently adopted strategy, which is organized around three imperatives: a modern transportation system, sufficient and equitable funding, and equity.[11] The plan provides a solid foundation for decision making and advances an important list of goals: economic opportunity, climate equity, facility preservation, maintenance, safety, innovative technologies, and congestion pricing. Just five months after the IIJA's enactment, ODOT had already made good on a community engagement goal.[12] Early feedback—assembled through a series of public meetings and online surveys—suggests the public would prioritize local street and school routes safety over better designed, maintained highways. The fresh strategy and a head start on outreach put Oregon in a good position to deploy its formula funds and win more than its fair share of competitive grants.

Broadband and water also are pressing needs. A 2020 Infrastructure white paper by the University of Oregon's Institute for Policy Research & Engagement concluded that "many rural and urban communities alike lack basic broadband infrastructure."[13] A 2016 report by the League of Oregon Cities, concluded that Oregon will need to invest $7.6 billion in water quality and supply infrastructure between 2016 and 2036: $4.3 billion for water quality projects and $3.3 billion for water supply projects.[14]

At the broadest level, the condition of Oregon's infrastructure is deteriorating. The American Society of Civil Engineers 2019 "Report Card for Oregon's Infrastructure" gave Oregon a C-minus grade with energy and wastewater systems receiving D grades.[15] In short, the state is not adequately financing infrastructure to meet deferred maintenance on top of capital needs to accommodate growth.

---

10   https://www.oregon.gov/gov/policies/Pages/infrastructure-investment-and-jobs-act.aspx

11   Oregon Department of Transportation (November 2021) *Strategic Action Plan*. ODOT. Salem, OR

12   Oregon Department of Transportation (March 2022) *Infrastructure Investment and Jobs Act STIP Update: Public Input Summary*. ODOT. Salem, OR

13   Oregon's Infrastructure Opportunities: Funding, Economic Development, and Resilience, Institute for Policy Research & Engagement, University of Oregon, October 2020.

14   *Infrastructure Survey Report (Water)*. (2016). League of Oregon Cities. https://www.orcities.org/application/files/9315/6115/9582/Water_Infrustracture_Survey_Report_Technical_FINAL7-29-16.pdf

15   American Society of Civil Engineers. (2019). Report Card for Oregon's Infrastructure.

Stepping Back: Infrastructure in the Context of Oregon's Economic Recovery

Transportation and utility stakeholders will naturally plan and evaluate potential IIJA investments through their individual lenses and promote safe, equitable, well-maintained facilities that deliver quality services to users. But to make the most of this once-in-a-generation opportunity, the Cabinet and other top elected and agency officials must also put the investment into the broader context of current conditions and the economic recovery.

Emerging from the pandemic, Oregon is a middling performer with many of the same natural assets as its West Coast neighbors - but with lower incomes and a less diversified economy. Like elsewhere, the Covid-19 downturn impacted communities of color, women, and low-income workers especially hard and exacerbated already wide inequities in wealth and income. The murders of George Floyd, Ahmaud Arbery, and countless other Black Americans sparked sustained protests in Portland and put a renewed focus on the present-day consequences of the State's unique 19th Century black exclusion law[16] and other systemically racist policies that followed.

Chronically undersupplied housing has triggered an affordability crisis that has put home buying out of reach for many, further exacerbated wealth inequities, and strained household budgets. The housing shortage is the principal driver of the state's growing and seemingly insoluble homelessness crisis. Housing affordability and homelessness are unrivaled, top public concerns.

Economic developers were recently stung by news that Intel, the state's largest private employer, chose Ohio for a sizable chip-making investment. A national imperative on domestic semiconductor manufacturing, and on re-shored supply chains generally, has Oregon officials re-examining their complex land-use and regulatory processes.

Roads must be safe, water clean, and broadband reliable. Those are necessary but insufficient aims of the IIJA. An infrastructure investment of this scale can and should do more. Implemented well, the IIJA should be a key ingredient to an inclusive recovery that puts Oregon on a path to more broadly shared prosperity. To achieve that, the IIJA should:

Rebuild and reconnect Oregon's economically isolated neighborhoods and communities

Harvard University economist Raj Chetty's research has demonstrated the importance of place on intergenerational economic mobility. Physical isolation is a major mobility barrier, and too many low-income children in Oregon and the

---

16 https://sos.oregon.gov/archives/exhibits/black-history/Pages/context/chronology.aspx

U.S. grow up in neighborhoods and schools with high poverty concentrations. Neighborhood and school segregation cuts off access to income-diverse networks, mentors, and jobs.

Oregon has a nationally relevant, rebuilding opportunity in Portland's Albina neighborhood. Hundreds of Black families were displaced, and homes, worth an estimated $1 billion today were destroyed to make way for Interstate-5 and subsequent 1960s era construction projects.[17] A decade-old initiative, the Albina Vision, seeks to develop a vibrant, 94-acre community within walking distance of downtown. The project enjoys strong political and philanthropic support but is complex and involves a dozen property owners. IIJA-funded investments—through both formula and competitive channels—could underwrite some or part of the initiative's concepts: a robust freeway cap, a rebuilt street grid, development of a waterfront park, and the realignment of intersecting light rail lines. The federal resources should take the initiative from blueprints into implementation.

The IIJA's broadband monies provide thousands of more opportunities to connect isolated households to jobs, education, and medical care across urban and rural Oregon. Federal resources are sufficient to significantly improve rates of connectivity, but the newly connected communities will rightly ask—connectively at what cost and to what end? The state's Broadband Advisory Council should develop a framework for adoption and make the case, community by community, on the importance of information communication technology.

Contribute to the need to build 30,000 new housing units annually for the next twenty years

Chronic housing underproduction has led to low housing vacancy rates, high prices and rents, and elevated levels of cost-burdening. A study ordered by the Legislature found that Oregon's housing stock is underbuilt by 140,000 units and, to address legacy underproduction and accommodate anticipated population growth, localities would need to build about 30,000 units annually through 2040.[18]

Local communities are quick to identify inadequate infrastructure as a top barrier to housing production. Hundreds of acres that have successfully transitioned into the state's constrained UGBs sit undeveloped because the lack of road, water, and sewer systems. And recent efforts to publicly subsidize affordable housing must divert dollars to build underlying infrastructure. So, an important question for IIJA investments: what can strategic infrastructure investment do to address the state's housing crisis?

---

17   https://www.oregonlive.com/commuting/2022/06/city-council-makes-informal-commitment-to-lower-albina-redevelopment-plan.html

18   https://www.oregon.gov/ohcs/about-us/Documents/RHNA/RHNA-Technical-Report.pdf

Oregon and its localities have dozens of systemic and project-level opportunities in the housing space, but the Instructure Cabinet and others will need to elevate housing as a strategic imperative. At the systemic level, architects of the state's emerging housing production legislation are considering state-centralized infrastructure funding to expedite land development. That could take the form of a system development charge (SDC) reimbursement fund—that could backfill SDC discounts that localities provide to local housing developers. Alternatively, the State could extend existing loan and grant programs that incent commercial and industrial development into the residential sector—as recently outlined in Business Oregon's recently released *Equitable Economic Recovery Plan*[19]. IIJA funds would provide a meaningful initial deposit into reimbursement, loan, or grant programs and accelerate the State's nascent production initiative.

Geographically focused opportunities exist across the State. Portland's 82nd Avenue, a former state highway recently transferred to city ownership, is a prime candidate for redevelopment as a major transit, pedestrian, and residential corridor. Long-idled industrial sites along the State's rivers are prime candidates for mixed-use development, including long discussed opportunities in Springfield and St. Helens. The Southeast UGB expansion area is primed for development but lacks road infrastructure. And the State should earmark some funds to support workforce housing in rural communities across Oregon.

Position Oregon to grow its manufacturing sector in an era of production reshoring

The pandemic disrupted global supply chains, underscoring the importance of maintaining a strong domestic manufacturing sector. Constrained supplies of the components of Covid-19 testing kits and personal protective equipment got the attention of elected leaders and public health officials. The disruption ushered in a new era of U.S. industrial policy and reshoring opportunities. If the public and private attention to domestic industrial persists, Oregon could grow manufacturing jobs rather than simply recover. A 10-20 thousand manufacturing job increase over the decade would be a reasonable aspiration if the Biden Administration and Congress maintain a focus on industrial policy.

Oregon has evolved into a manufacturing leader during the past half century and now has an opportunity to continue to lead and grow equitably in a new era of policy and investment. Developers and business leaders, however, have long suggested that Oregon has an inadequate supply of project-ready sites for industrial and other uses. Like the housing issue, preliminary analysis indicates that inadequate infrastructure stands in the way of business growth because of an adequate inventory of project-ready sites.

---

19 ECONorthwest (March 2022) *Equitable Economic Recovery Plan.* Prepared for Business Oregon. Salem, OR. Page 41

Bend's 500-acre Juniper Ridge site underscores the critical role of infrastructure in economic development. The land, targeted for industrial uses and business parks, successfully transitioned into the UGB in 2003. More than a decade later, the site sat empty and required a major sewer line. The city completed the first two phases of the project during 2020-2021, and timing of the third, final stage is to-be-determined.[20] It will have taken at least 20 years for the site to transition from greenfield to developable lot. IIJA funding could be instrumental in helping Bend bring Juniper Ridge across the finish line while also strengthening the competitiveness of dozens of other partially developed sites.

Conclusion

Based on the composition of its funding, IIJA has been described as an infrastructure act with a transportation focus. The Act's proponents clearly intended that newly supported projects create near-term construction jobs and demonstrate that a divided Congress can come together on issues of national importance. But Oregon would be well served if the vision for the IJJA were even more ambitious. Rather than a job-creating road, water, and broadband bill, Oregon stakeholders should view the Act as a critical building block for three of the State's most pressing economic issues: equity, housing, and industrial revitalization. A shared vision around these imperatives together with well-coordinated, high-return investments would boost Oregon's economic competitiveness and set the stage for more broadly shared prosperity.

Author Capsule Bios

Robert Parker is Director of Strategy and Technical Solutions for the University of Oregon's Institute for Policy Research and Engagement (IPRE). Since 1989, he has managed more than 500 policy and planning analysis projects with communities and state agencies throughout Oregon. He served as Executive Director of IPRE from 2000-2020. IPRE is known widely throughout Oregon as one of the state's critical policy analysis resources, connecting expertise of University faculty and students with communities and agencies. These relationships, as well as the vast policy analysis experience, help IPRE transform Oregon communities and organizations through research and action. He also coordinates the EDA University Center for economic development housed in IPRE.

John Tapogna is a Senior Policy Advisor at ECONorthwest with a focus on economic development, fiscal, housing, workforce, education, and social policy. Since his arrival in 1997, his clients have included the Bill and Melinda Gates Founda-

---

20   https://www.bendoregon.gov/city-projects/infrastructure-projects/north-interceptor

tion, the Lumina Foundation, the Oregon Business Council, and numerous state agencies and local governments across the United States. He served as ECONorthwest's president from 2009-2021 and oversaw a strategy that successfully transitioned the firm to a second generation of ownership and expanded its presence along the West Coast and into the Intermountain West. Prior to joining ECONorthwest, John was an Analyst at the U.S. Congressional Budget Office and a Peace Corps volunteer in Chile.

Winter Storm Uri: Resource Loss and Psychosocial Outcomes of Critical Infrastructure Failure in Texas

Liesel A. Ritchie,[1,2] Duane A. Gill,[3] Kathryn Hamilton[4]

[1] Professor of Sociology and Associate Director, Center for Coastal Studies at Virginia Polytechnic Institute and State University

[2] Corresponding Author, liesel14@vt.edu

[3] Research Professor of Sociology at Virginia Polytechnic Institute and State University

[4] Graduate Student, Department of Sociology at Virginia Polytechnic Institute and State University

Abstract

In February of 2021, Winter Storm Uri affected parts of the United States, Mexico, and Canada. Texas was particularly hard hit, as the state's primary power provider, ERCOT (the Electric Reliability Council of Texas), proved to be unprepared for the event—despite similar storms in 1989 and 2011 that revealed weaknesses in the state's electric grid system. This article investigates psychosocial outcomes of individuals who experienced Winter Storm Uri. Drawing upon survey data collected in Texas in April and May of 2022, we illustrate ways in which loss of critical infrastructure and compounding results influence levels of stress among respondents. Using Hofoll's (1989, 1991) Conservation of Resources model of stress, we find that Uri-related losses of objects and conditions resources contribute to elevated stress as measured by the Avoidance subscale of the Impact of Event Scale (Horowitz 1976; Horowitz, Wilner, and Alvarez 1979)— more than one year after the disaster. Our regression model consisting of indicators of objects resource loss, conditions resource loss, and demographic characteristics explains approximately 33 percent of the variance in the Avoidance subscale. Findings suggest that more attention should be paid to the social impacts of critical infrastructure failures and that such impacts should be addressed by improving critical infrastructure policy and regulations, as well as the physical structures.

Keywords: Conservation of Resources Model, Impact of Event Scale, Psychosocial Stress, Winter Storm Uri, ERCOT, Critical Infrastructure

Introduction

The February 2021 North American winter storm—also known as Winter Storm Uri—was an intense ice storm that ravaged parts of the United States, Mexico, and Canada, leaving devastation in its path. The storm began in the Pacific Northwest and rapidly advanced to the Midwestern and Northeastern regions of the United States. The National Weather Service issued winter weather alerts to millions of Americans. In the U.S., Uri caused blackouts for nearly ten million people and, most notably, was the root of the 2021 Texas power crisis.

As noted by Popik and Humphreys (2021), Texas maintains its own electric grid managed by the Electric Reliability Council of Texas (ERCOT). This grid is separate from the nation's other systems, a move intended to reduce the cost of power by eliminating federal tariffs and regulation of transmission lines. Despite experiences associated with extreme cold weather conditions in 1989, 2011, and 2014 that demonstrated that ERCOT was ill-prepared to handle such situations, little was done to weatherize plants or to plan for future power outages (Popik and Humphreys 2021; Vogel and Vogel 2021). As stated by Vogel and Vogel (2021):

> Ultimately, the Texas authorities' faith in their markets reinforced their propensities not to plan, coordinate, and invest; to favor low costs over resilience; and to privilege industry profits over the protection of retail consumers. Yet the market design flaws were not so fundamental that the system could not have avoided massive blackouts with a little bit of good old-fashioned planning, coordination, regulation, and public investment.

Consequently, Winter Storm Uri caused a major power grid failure in Texas, resulting in water, heat, and food shortages (Hobby School of Public Affairs 2021).

On February 15, 2021—the height of the blackouts around the state—roughly one-third of ERCOT's customers had no electricity (Popik and Humphreys 2021). According to a research report by the University of Houston's Hobby School of Public Affairs (2021), more than 4.5 million homes were without power for several days; approximately 69 percent of Texans lost electricity for an average of 42 hours at some point during the storm; and almost one-third of those who responded to an on-line survey indicated they had water damage to their homes (Stipes 2021). Death toll estimates range from 246 (Texas Department of State Health Services 2021) to more than 700 (Aldhous, Lee, and Hirji 2021), depending on the source. In total, Winter Storm Uri left an estimated $295 billion in damages in its wake (Stipes 2021). Yet, while residents of Texas suffered during the crisis, some energy firms made billions through an increase of wholesale prices (Vogel and Vogel 2021).

Although there are numerous peer reviewed articles and reports reviewing the technical causes, policy drivers, and consequences of the power grid failure resulting from Winter Storm Uri (e.g., see Glazer at al.; Li et al. 2022; Nejat et al. 2022; Popik and Humphreys 2021; Smead 2021; Vogel and Vogel 2021), there is surprisingly limited published information on the social dimensions of the event in Texas (for exceptions see Bottner et al. 2021; Hobby School of Public Affairs 2021; Li et al. 2022). In this article, we address the following research question: *What is the relationship between critical infrastructure failure and psychosocial stress in the context of Winter Storm Uri?* To attend to our research question, we rely on Hobfoll's (1989, 1991) Conservation of Resources (COR) theory and characterize critical infrastructure failure as a type of resource loss (Hobfoll 1989, 1991). In our analysis, we operationalize stress using the Avoidance subscale of the Impact of Event Scale (Horowitz 1976; Horowitz, Wilner, and Alvarez 1979).

Theoretical Framework

Typically, studies using COR theory measure losses and gains of different types of resources related to a particular event or situation (e.g., see Binder et al. 2020; Gill, Picou, and Ritchie 2012; Gill et al. 2014; Ritchie, Gill, and Long 2020; Ritchie, Little, and Campbell 2018). The COR model of psychosocial stress employs four types of resources that people value: (1) objects, such as physical possessions, and natural and built environments; (2) conditions, such as good interpersonal relationships, employment, stable living arrangements, or other stabilizing social circumstances; (3) personal characteristics, including self-efficacy, self-esteem, and sense of optimism; and (4) energies, among which are time, financial savings, and knowledge (Hobfoll 1989, 1991).

In this article we focus on objects resource losses and conditions resource losses associated with Winter Storm Uri. Among the former are damages to respondents' residences or neighborhoods, financial losses, and loss of access to various services and modes of communication. The latter, conditions resources, include experiences with the storm, storm-related social disruption, and personal serious injury or serious injury of someone close to survey respondents.

The bases of the COR theory are that stress is caused when people experience resource loss, threat of resource loss, or invest resources without return on such investments (Hobfoll 1989, 1991). This is especially the case when losses are sudden, as is the case in most disasters, or when disasters produce impacts that threaten further losses. "Resource caravans," which represent the combination of the four different types of resources mentioned above (Hobfoll 2011a, 2011b), have the potential to either temper resource losses or to lead to resource loss spirals or "stalled resources," particularly in the context of distinct events (Halbesleben et al. 2014; Hobfoll 1991). Although disasters may offer some opportunities for resource gains among some groups, this is rare (e.g., see Tierney 2014).

As summarized in work by Halbesleben and colleagues (2014), the impacts of resource loss are greater than resource gain. Moreover, acquisition of resources (resource gain) necessitates resource investment; investments tend to provide a buffer against resource loss and support recovery from resource loss. Additional research (Halbesleben 2010; Halbesleben and Bowler 2007; Mäkikangas, Bakker, Aunola, and Demerouti 2010) shows that individuals who have limited resources are more likely than those who have more abundant resources to experience resource losses (Benight, Swift, Sanger, Smith, and Zeppelin 1999; Halbesleben and Bowler 2007; Hobfoll 2001). Finally, where losses are widespread, community-level stress has a tendency to create what Hobfoll (1991) refers to as a "pressure cooker effect," which can result in diminished social capital (Ritchie 2012). Clearly, COR theory is applicable to disaster studies and has increased in use since its inception (Arata, Picou, Johnson, and McNally 2000; Benight et al. 2007; Binder et al. 2020; Campbell, Vickery, and Ritchie 2018; Clay and Greer 2019; Gill and Ritchie 2018; Ritchie 2012; Sattler, Whippy, Graham, and Johnson 2018). Thus, we employ COR theory in our research on Winter Storm Uri.

Methods

We created our survey instrument using or adapting existing items from previous post-disaster research (e.g., see Binder et al. 2020; Gill, Picou, and Ritchie 2012; Gill et al. 2014; Ritchie, Gill, and Long 2020; Ritchie, Little, and Campbell 2018). Our questions captured information about critical infrastructure failures and damages; storm experiences; evacuation behaviors; social disruption; and other factors hypothesized to influence stress avoidance behaviors. The survey also included a module of questions from the Impact of Event Scale Avoidance subscale, described in detail in the Analytic Approach and Results section, below.

Qualtrics, a survey research firm, administered the on-line survey using a panel, quota sampling approach. This helped to ensure that the final sample reflected Texas demographics (e.g., see Evensen et al. 2017). Eligible participants were Texas residents living in the state at the time of the storm, ages 18 and older. We field tested the instrument with 65 individuals, which resulted in minor revisions (deletions) of some response categories that yielded little to no data. Data collection took place in April and May of 2022. Upon quality checking, the total number of completed surveys was 1,567. The median time for survey completion was 17 minutes. Of the 1,567 respondents, this article focuses only on those who indicated they resided in an area that was affected by the storm.

Subsample Characteristics

A total of 1,295 respondents (82.7%) indicated they resided in an area that was adversely affected by Winter Storm Uri. Within this subsample, there was an even split between females (50.1%) and males (49.9%), and 70 percent identified as white, which is comparable to the state of Texas as a whole (50.3% female and 79% white) (U.S. Census 2020). The subsample was higher educated than the state as a whole with 96 percent of respondents aged 25 or older reporting having at least a high school degree compared to 84 percent of the state population of similar age. In terms of annual household income before taxes, slightly more than one-fourth of the subsample reported less than $35,000 and about one-third reported more than $100,000. The subsample median household income was in the $50,000 to $74,999 range, which was comparable to the state's $63,826 median household income.

In terms of household composition, the median household size for respondents was 3.1 persons compared to the state average of 2.8 persons per household. Almost 45 percent of the households had children under the age of 18 and one-third had disabled or elderly household members. About two-thirds (64.2%) owned their residences, which is close to the state percentage of 62.3. Almost 45 percent of the group had lived in their neighborhood for 10 years or more.

Analytic Approach and Results

Our analysis focuses on relationships between avoidance behaviors and background characteristics, loss of access to various services and modes of communication, physical damages, storm experiences, social disruption, and other forms of resource loss. We begin with a description of the Avoidance subscale as our dependent variable. We then proceed with bivariate analysis examining the subscale's relationship with background characteristics, objects resource losses, and conditions resource losses. This is followed by a multivariate analysis using Ordinary Least Squares (OLS) regression.

Dependent Variable: The Avoidance Subscale

The Avoidance subscale from the Impact of Event Scale (IES) is our measure of psychosocial stress. Horowitz and colleagues developed the IES as a self-reported indicator of posttraumatic stress employing the rationale that highly stressful events are likely to elicit intrusive thoughts and feelings, as well as efforts to cope with these intrusions by avoiding reminders of the traumatic event (Horowitz 1976; Horowitz, Wilner, and Alvarez 1979). The IES is anchored in a specified event asking, "Thinking of the (event) please indicate how often each one (item) was true for you during the past seven days." There are 15 items measured on a four-point scale to record how often each item was experienced during the previous seven days (0 = not at all; 1 = rarely; 3 = sometimes; 5 = often). The following eight items comprise the Avoidance subscale: I had to stop myself from getting upset when I thought about it; I tried to remove it from my memory; My feelings about it were kind of numb; I had a lot of feelings about it that I didn't know how to deal with; I stayed away from reminders of it; I felt as if it had not really happened; I tried not to talk about it; and I tried not to think about it. Avoidance behaviors are of particular importance from a sociological perspective because such behaviors tend to adversely affect communication, social interactions, and social networks, which in turn, weakens social connections, reduces trust, and diminishes social capital.

The IES and its subscales have been used in several studies of technological disasters. Green and colleagues (1994) used it in their longitudinal study of child survivors of the 1972 Buffalo Creek dam collapse and flood, and Davidson and Baum (1986) used the IES in their study of the 1978 Three Mile Island nuclear disaster. More recently, the scale and subscales were used as measures of initial and chronic psychosocial stress associated with the 1989 *Exxon Valdez* oil spill (Gill, Ritchie and Picou 2016), the 2010 BP *Deepwater Horizon* oil spill (Gill, Picou and Ritchie 2012; Gill et al. 2014; Ritchie, Gill and Long 2018), and the 2008 Kingston, TN coal ash spill (Ritchie, Little and Campbell 2018; Ritchie, Gill and Long 2020; Ritchie and Long 2021). Most of these studies directly or indirectly linked various types of resource loss with higher IES and subscale scores.

Analysis of our Uri data reveals a high degree of reliability for the Avoidance subscale with an alpha of .90. The mean for the Avoidance subscale is 11.6 for the subsample, which is comparable to Avoidance subscale means of 11.3 for a sample of south Mobile County residents five months after the BP *Deepwater Horizon* oil spill and 11.0 for a sample of Cordova residents five months after the *Exxon Valdez* oil spill. Extrapolating from clinical scoring of the IES, scores on the Avoidance subscale indicate that approximately 16 percent of subsample respondents are in the 'severe' clinical range 14 months after Winter Storm Uri.

Background Characteristics and Avoidance Behaviors

Drawing on subsample characteristics previously described, we examined relationships between them and avoidance subscale scores. As shown in Table 1, significantly higher levels of avoidance behaviors were reported for female and non-white respondents compared to their counterparts. Respondents from households with children under the age of 18 and those with disabled or elderly household members had significantly higher avoidance scores relative to those who did not have those types of dependents. Younger respondents and those with lower household incomes reported significantly higher levels of avoidance than older respondents and those with higher household incomes. Residential owners and those who had lived longer in the community reported lower levels of avoidance than those who rented and were more recent community members.

Objects Resource Loss: Critical Infrastructure and Services

One type of resource involves access to critical infrastructure such as utilities and means of communication, as well as access to various goods and services such as groceries, gasoline, banking, medical and the like. It is expected that respondents

Table 1. Descriptive and Bivariate Statistics for Background Characteristics and the Avoidance Subscale for Winter Storm Uri Impact Group

Background Variable	N	Percentages		Avoidance t value
Gender	1289	Female = 50.1	Male = 49.9	3.29***
Race	1295	White = 70.3	Non-white = 29.7	-3.82***
Marital Status	1283	Married/partner = 61.7	Not married = 38.3	-1.55
Residential Status	1295	Own = 64.2	Rent = 35.8	3.17**

	N	Mean	Avoidance Correlation (r)
Age[a]	1289	40-44 years	-.285***
Education[a]	1293	Some College	.005
Household Income[a]	1295	$50,000 -$74,999	-.099***
Years in Community[a]	1295	4-6 years	-.065*
Household size	1289	3.15	.171***
Household members under 18	1289	0.84	.208***
Disabled or Elderly Household members	1289	0.53	.096***

[a] Variables were measured using ordinal categories: age (18-24 = 11.4%; 25-29 = 8.6%; 30-34 = 13.6%; 35-39 = 9.5%; 40-44 = 9.3%; 45-49 = 8.4%; 50-54 = 6.6%; 55-59 = 8.1%; 60-64 = 5.9%; 65+ = 18.8%); education (less than 9th grade = 0.5%; some high school = 3.1%; high school diploma = 19.0%; some college or vocational school = 33.7%; BA or BS degree = 22.8%; some graduate work = 3.4%; advance degree = 17.4%); household income (less than $10,000 = 6.3%; $10,000 to $14,999 = 3.7%; $15,000 to $24,999 = 7.8%; $25,000 to $34,999 = 8.6%; $35,000 to $49,999 = 10.2%; $50,000 to $74,999 = 19.5%; $75,000 to $99,999 = 11.5%; $100,000 to $149,999 = 20.8%; $150,000 to $199,999 = 7.6%; $200,000 or more = 3.9%); and years in the community (0-3 = 24.6%; 4-6 = 19.0%; 7-10 = 11.9%; more than 10 years = 44.5%).
* $p < .05$ level; ** $p < .01$ level; *** $p < .001$ level (two-tailed)

Table 2. Loss of Access to Utilities, Communication, and Services and their Relationship to the Avoidance Subscale for Winter Storm Uri Impact Group

Type of Service	Length of time without access (percent)						Correlation with Avoidance Subscale Coefficient
	Never had or lost access	1-12 hours	13-24 hours	2-3 days	4-7 days	Longer than 1 week	
Utilities							
Electricity	15.4	19.2	15.8	27.7	17.2	4.8	.223**
Water	38.4	12.8	9.5	19.8	13.5	5.9	.219**
Natural Gas	71.7	5.7	5.4	8.7	5.5	3.0	.401**
Communication							
Television	21.0	19.9	13.9	24.7	15.8	4.7	.237**
Radio	54.1	11.4	8.7	13.7	9.3	2.8	.258**
Internet	23.4	20.5	13.4	22.1	14.9	5.7	.250**
Land line phone	68.4	7.0	6.0	9.0	6.7	2.9	.307**
Cell Phone	67.6	11.2	7.3	7.7	4.4	1.7	.319**
Text messaging	67.7	11.8	6.7	8.0	3.9	1.8	.331**
Social Media	60.8	13.4	8.2	9.7	5.5	2.4	.297**
Services							
ATMs	47.5	7.5	9.4	19.1	12.1	4.3	.222**
Banks	39.5	6.0	11.1	23.6	14.2	5.6	.191**
Gas Stations	38.5	11.3	13.3	24.5	9.9	2.6	.197**
Grocery stores	29.2	10.9	15.5	27.5	12.9	3.9	.196**
Pharmacies	38.4	9.9	12.8	22.9	12.6	3.4	.194**
Hospitals/medical facilities	62.8	5.6	7.2	13.5	8.3	2.7	.169**
Transportation	54.3	5.7	7.3	17.5	10.4	4.7	.254**
Place of work	44.2	4.2	6.9	21.4	16.1	7.2	.210**
Schools	42.0	3.8	6.1	19.3	19.7	9.0	.238**
Childcare	73.9	2.9	4.6	8.4	6.7	3.5	.288**

** $p < .01$ (two-tailed)

who experienced longer periods of lost access and those who suffered physical damages and losses will have greater levels of psychosocial stress as measured by the Avoidance subscale.

Respondents were asked to indicate how long they were without access to various goods and services—never had access or never lost access (= 0), 1-12 hours (= 1), 13-24 hours (= 2), 2-3 days (= 3), 4-7 days (= 4), or longer than one week (= 5). Table 2 shows the results. For utilities, almost one-half of respondents indicated a loss of electricity and 40 percent were without water for two days or more. An examination of access to various forms of communication reveals that approximately two-thirds of respondents never had or lost access to cell phones, land line phones, text messaging, and social media, while more than 40 percent lost access to TV and the Internet for two days or more. Almost one-half (48%) of the respondents lost access to schools for two days or more and about four out of 10 respondents lost access to their place of work (45%), grocery stores (44%), banks (43%), pharmacies (39%), and gas stations (37%) for two days or more. Loss of access to each type of utility, form of communication, and category of service was significantly correlated with the Avoidance subscale.

A Utilities Access Loss scale was created by combining responses to length of loss access to electricity, water, and natural gas. This resulted in a scale with a range of 0 to 15, a mean of 4.7, a standard deviation of 3.5, and an alpha of .63, which was significantly correlated with the Avoidance subscale (.36; $p < .01$). A Communication Access Loss scale was created by combining responses to television, radio, internet, land line phone, cell phone, text messaging, and social media. This resulted in a scale with a range of 0 to 35, a mean of 8.6, a standard deviation of 7.6, and an alpha of .88, which was significantly correlated with the Avoidance subscale (.37; $p < .01$). Finally, a Services Access Loss scale was created by combining service items (gas stations, banks, ATMs, grocery stores, pharmacies, hospital/medical facilities, transportation, schools, childcare, and place of work), which resulted in a scale with a range of 0 – 50, a mean of 15.8, a standard deviation of 12.1, and an alpha of .90, which was significantly correlated with the Avoidance subscale (.27; $p < .01$). These scales were used in the regression analysis.

Objects Resource Loss: Property Damage

Respondents reported property damage by answering a series of questions where yes = 1 and no = 2. As shown in Table 3, damages include residential, other property, neighborhood/community, and financial losses. More than one-third of respondents (36%) reported damages to their place of residence and 29 percent indicated they experienced other property damage. A majority (62%) observed damages to their neighborhood and 46 percent experienced financial losses. T-test results indicted significant relationships ($p<.001$) between these forms of property damage/loss and the Avoidance subscale.

Table 3. Descriptive and Bivariate Statistics for Reported Property Damage and the Avoidance Subscale for Winter Storm Uri Impact Group

Type of Loss	Percent		Avoidance
	Yes	No	t value
Residential Damage	36.1	63.9	10.66***
Other Property Damage	28.7	71.3	6.59***
Neighborhood/Community Damage	61.7	38.3	8.41***
Financial Losses	45.9	54.1	8.55***

*** $p < .001$ level (two-tailed)

A Physical Damages scale was created by adding the following variables: residential damage, other property damage, neighborhood damage, and financial losses. The scale was recoded (0 = no loss to 4 = high loss) and had a mean of 1.7, a standard deviation of 1.4, an alpha of .71, and was significantly correlated with the Avoidance subscale (.32; $p < .01$). This scale was used in the regression analysis.

Conditions Resource Loss: Storm Experiences, Social Disruption, and Serious Injuries

Conditions resource loss consisted of storm experiences, social disruption, and serious injuries. As measures of resources in the context of COR theory, these variables are indicators of threats to stable conditions of daily life. For example, fear associated with storm experiences threatens stability, as do social disruption and serious injuries. It is expected that severe storm experiences, high levels of social disruption, and serious injuries to oneself or loved ones will contribute to increased levels of psychosocial stress as measured by the Avoidance subscale.

Storm Experiences

Three items served as indicators for storm experiences: "to what extent did you feel safe during the storm?" (1 = perfectly safe – 5 = life-threatening danger); "to what extent did you feel afraid during the storm? (1 = not at all afraid – 5 = very afraid); and "during the storm, I felt unable to control the important things in my life (1= strongly disagree – 5 = strongly agree). As shown in Table 4, almost 40 percent of respondents reported life-threatening or near life-threatening feelings and being afraid or very afraid. More than 40 percent agreed or strongly agreed that they felt unable to control the important things in life. All three indicators were significantly correlated with the Avoidance subscale.

A Storm Experience Scale was created by adding the three indicators. The scale ranged from 3 to 15, had a mean of 9.1, a standard deviation of 2.9, an alpha of .70, and was significantly correlated with the Avoidance subscale (.44; p <.001). This scale was used in the regression analysis.

Table 4. Descriptive and Bivariate Statistics for Storm Experiences and the Avoidance Subscale for Winter Storm Uri Impact Group

Experience	Response (Percent)					Avoidance Subscale Correlation
To what extent did you feel safe?	Perfectly safe 11.0	20.1	31.3	28.6	Life-threatening 9.0	.234***
To what extent did you feel afraid?	Not at all afraid 17.1	17.8	26.6	28.4	Very afraid 10.1	.437***
I felt unable to control the important things in my life	Strongly disagree 14.9	17.1	25.4	26.5	Strongly agree 16.1	.372***

*** $p < .001$ level (two-tailed)

Social Disruption

Social disruption was measured by asking respondents the extent to which they agreed or disagreed with statements indicating disruptions in the household, community, and the State of Texas, as well as statements regarding resolution of impacts at these three levels. Responses were scored 1 = strongly disagree; 2 = disagree; 3 = neither; 4 = agree; and 5 = strongly agree. Results found in Table 5 indicate that more than two-thirds of the respondents agreed or strongly agreed that the storm disrupted their household (69%) and community (68%). Eight out of ten respondents agreed that the state of Texas experienced social disruption. At the same time, the vast majority agreed or strongly agreed that storm impacts had been resolved for their household (77%) and community (72%), but less than half (43%) thought that impacts to the state of Texas had been resolved. All individual items were significantly correlated with the Avoidance subscale.

A Social Disruption Scale was computed by adding the six indicators. The scale had a range of 6 to 30, a mean of 22.8, a standard deviation of 3.7, an alpha of .62, and was significantly correlated with the Avoidance subscale (.09; p<.001). This scale was used in the regression analysis.

Serious Injuries

Respondents reported property serious injuries by answering a series of questions where yes = 1 and no = 2. Serious injuries included those experienced by the respondent and those experienced by someone close to the respondent. More than

Table 5. Descriptive and Bivariate Statistics for Social Disruption and the Avoidance Subscale for Winter Storm Uri Impact Group

Level of Disruption	Response (Percent)					Avoidance Subscale Correlation
	Strongly Disagree	Disagree	Neither	Agree	Strongly Agree	
The impacts of the storm caused disruption in my household	6.9	8.6	15.1	39.0	30.4	.163***
The impacts of the storm caused social disruption in my community	6.0	7.7	18.1	38.3	30.1	.201***
The impacts of the storm caused social disruption in the State of Texas	3.8	3.6	11.8	33.7	47.2	.129***
The impacts of the storm on my household have been resolved	1.8	5.8	15.1	46.5	30.8	-.171***
The impacts of the storm on my community have been resolved	2.1	6.9	19.3	46.1	25.6	-.148***
The impacts of the storm on the State of Texas have been resolved	8.8	18.6	29.4%	29.7	13.5	.073***

*** p < .001 level (two-tailed)

Table 6. Unstandardized (b) and Standardized (β) Regression Coefficients and Standard Errors (SE) for Determinants of Avoidance Behaviors for Winter Storm Uri Impact Group

	Avoidance Subscale			
Independent Variables	b	β	SE	Sig
Objects Resource Loss Variables				
Utility Access Loss Scale	.26	.09	.10	.009
Communication Access Loss Scale	.14	.16	.05	.003
Services Access Loss Scale	.01	.01	.03	.667
Property Damage Scale	.87	.10	.05	.000
Conditions Resource Loss Variables				
Storm Experience Scale	.93	.26	.10	.000
Serious Injury Scale	2.32	.10	.66	.000
Social Disruption Scale	.156	.06	.07	.024
Demographic and Control Variables				
Age	-.52	-.15	.10	.000
Gender (Female or Male)	.39	.02	.55	.479
Race (White or Non-White)	1.67	.07	.58	.004
Married/Partner or Not Married/Partner	.53	.03	.56	.342
Education	.38	.05	.21	.071
Household Income	-.22	-.05	.14	.107
Household Size	-.19	-.03	.28	.502
Number of Dependent Children	.70	.08	.32	.027
Number of Disabled or Elderly	.53	.04	.32	.099
Number of Years Lived in Community	.15	.02	.22	.502
Own or Rent?	.44	.02	.62	.482
Constant	-5.42		3.28	.098
Adjusted R^2	.325			
N	1200			

90 percent of the respondents escaped serious injury to themselves and to someone close to them. As expected, these were significantly related to the Avoidance subscale.

A Serious Injury scale was created by recoding two variables—personal serious injury and serious injury to someone close to you—where yes = 1 and no = 0, then adding the two. This resulted in a scale with a range of 0-2, a mean of 0.1, a standard deviation of 0.4, and an alpha of .62, which was significantly correlated with the Avoidance subscale (.29; p <.01). This scale was used in the regression analysis.

Regression Analysis

Next, we estimated a series of regression models to examine the effects of the independent variables affecting avoidance behaviors, controlling for alternative explanations and socio-demographic characteristics. We evaluated three sets of predictors (indicators of objects resource loss, conditions resource loss, and demographic variables) on the Avoidance subscale. All models were tested for multicollinearity, and no problems were detected. All VIF values were under 2.4.

As shown in Table 6, two of the three objects resource loss variables—utilities access loss and communication access loss—were found to have positive effects on the Avoidance subscale (p\.009 and p\.003 respectively). All three conditions resource loss variables—storm experiences (p\.000), serious injuries (p\.000), and social disruption (p\.024) —were positively related to avoidance. With the exception of age (p\.000), race (p\.004), and number of dependent children in the household (p\.027), the demographic and control variables had minimal impact on the Avoidance subscale. The adjusted R^2 values indicate a good model fit with 33 percent of the variance in the Avoidance subscale explained by the combined predictors.

Discussion

Based on findings of a 2022 on-line household survey conducted in Texas, this article addresses the research question, *"what is the relationship between critical infrastructure failure and psychosocial stress in the context of Winter Storm Uri?"* Using the Conservation of Resources framework, we have shed light on ways in which resource losses associated with the storm influence levels of stress within a subsample of the population residing in an area adversely affected by Winter Storm Uri, as measured by the Avoidance subscale of the Impact of Event Scale. Results presented in this article are based on preliminary analyses of our quantitative data and further investigation is under way. With that said, the independent variables in our regression model explain approximately 33 percent of the variance in the Avoidance subscale, based on the Adjusted R^2.

In particular, we find that objects losses such as loss of utilities; loss of access to forms of communication; damages to homes, other property damages, and neighborhood damages; as well as financial losses related to critical infrastructure failure tend to heighten psychosocial stress. Results also show that conditions resource losses such as storm experiences, social disruption, and serious personal injuries or serious injuries to someone close to respondents can heighten avoidance behaviors.

Overall, as found in previous disaster studies employing the IES (e.g., see Gill, Picou, and Ritchie 2012; Gill et al. 2014; Ritchie, Gill, and Long 2020; Ritchie, Little, and Campbell 2018), there is a direct, positive relationship between objects and conditions resource losses and psychosocial stress. Although this is not surprising, it is notable that survey findings from more than one year after the storm reveal that approximately 16 percent of respondents reported engaging in clinical levels of avoidance behaviors as measured by the IES. Although some avoidance behaviors might be considered healthy coping mechanisms and positive strategies for moving on from impacts of the disaster, we argue that avoidance behaviors among this proportion of the population where individuals are withdrawn are not healthy. From a sociological perspective, such behaviors have the potential to diminish social capital by decreasing social interaction (Ritchie 2012).

Although males and females are significantly different with respect to avoidance behaviors in the bivariate analysis, gender becomes insignificant when other factors are considered in the regression model. Notably, younger respondents and respondents who reported loss of communication services were more likely than others to report engaging in avoidance behaviors. We suspect that loss of access to various forms of communication among this population reduces social support and accentuates avoidance behaviors. In the context of the COVID-19 pandemic, we speculate that the need for access to a variety of communication modes was of particular importance to most respondents. For those used to having and using such access—like younger adults—loss of it and diminished social interaction could contribute to increased avoidance behaviors and other forms of psychosocial stress.

The relationship between age and avoidance behaviors is an interesting one that warrants further examination. It suggests that a population not usually defined as "vulnerable"—adults between the ages of 18 and 45—might be vulnerable in some disaster situations. We refer to this phenomenon as a "new vulnerable" population, much like empirical research has shown in prior technological disasters. For example, research conducted in the wake of the *Exxon Valdez* and BP *Deepwater Horizon* oil spills revealed that those dependent upon renewable natural resources for their livelihoods and way of life were more likely to experience psychosocial stress than those who did not.

Future analyses will explore other factors that have been shown to affect psychosocial stress, including involvement in compensation processes, as well as perceptions of recreancy associated with issues of preparedness, response, recovery, and mitigation associated with Uri.

Conclusions

The impacts of Winter Storm Uri highlight the importance of critical physical infrastructure resilience—as well as social resilience—as recommended by the National Institute of Standards and Technology in its Community Resilience Planning Guide (2016). The keys to developing and maintaining physical infrastructure include identifying and characterizing the built environment. Similarly, social resilience necessitates characterizing social functions and dependencies and linking these with the built environment (National Institute of Standards and Technology 2016). This did not happen in the case of Uri and seems not to be happening now with ERCOT's lack of preparedness for power grid demands as the summer of 2022 approaches (Douglas and Ferman 2021; Morehouse 2022). Some clues as to the implications of this lack of preparedness may be found in our study.

Although data for this study were collected at the individual level, there are broader sociocultural implications. The situation in Texas is certainly worth monitoring as the unpredictability of ERCOT's ability to provide its constituents with the power they need to live and function on a daily basis continues to persist. Given the importance of water in generating electrical power, increasing temperatures and drought conditions strain many power-generating plants by limiting the supply of water, increasing the stress on equipment (warmer water increases the stress), and interfering with routine and needed maintenance. Hotter, drier weather this summer and future summers will very likely increase power plants failures as demand for electricity increases (Morehouse 2022). Already, the Texas grid operator has called on customers to conserve as much power as possible as the state has experienced a hard-hitting combination of record-breaking heat and high demand, coupled with low wind generation and the simultaneous failure of six thermal generators (Morehouse 2022). Ensuring there are enough resources to create power does not solve the entire problem and leaves the reliability of the power plants up to question (Douglas and Ferman 2021).

For those experiencing Uri-related stress, ongoing power outages can be constant reminders of the storm's impacts and the precarious context of their personal, household, and community wellbeing. As of the writing of this article (June 2022) ERCOT is experiencing current outages and is anticipating more throughout the summer of 2022 (Douglas and Ferman 2021; Morehouse 2022). The chronic nature of these circumstances has the potential to foment uncertainty and stress. At the same time, substantial increases in energy costs (Vogel and Vogel 2021) are likely to contribute to additional psychosocial stress and other adverse outcomes for community wellbeing. Given the high percentages of our respondents whose

Uri-related avoidance behaviors are at the clinical level and persisting over time, we contend that there is an important need for professional counseling that is going unmet or is not being sought, for whatever reasons. Policy- and decision-makers must start to take this into account as they work to address Texas's critical infrastructure—one that will both meet energy needs and the need to support the state's critical social infrastructure. Other states should take note and learn from the unfortunate example provided by Winter Storm Uri and ineffective human decision-making.

Acknowledgements

The authors would like to thank the College of Liberal Arts and Human Sciences and the Department of Sociology at Virginia Polytechnic Institute and State University for funding this research. We also extend our appreciation to Dr. Michael Edelstein and Dr. Michael Long for their feedback on our survey instrument, and to Dr. Long for his comments on early drafts of this manuscript.

Author Capsule Bios

Liesel A. Ritchie is a Professor of Sociology and Associate Director of the Center for Coastal Studies at Virginia Polytechnic Institute and State University. During her career, Dr. Ritchie has studied a range of disaster events, including the *Exxon Valdez* and BP *Deepwater Horizon* oil spills; the Tennessee Valley Authority coal ash release; Hurricane Katrina; earthquakes in Haiti and New Zealand; and Winter Strom Uri in Texas. Since 2000, her focus has been on the social impacts of disasters and community resilience, including conducting social impact assessments, with an emphasis on technological hazards and disasters, social capital, and rural renewable resource communities.

Duane A. Gill is a Research Professor of Sociology at Virginia Polytechnic Institute and State University. Dr. Gill is a disaster scholar specializing in technological hazards and disasters and has conducted extensive research on the 1989 Exxon Valdez oil spill, Hurricane Katrina, and the 2010 BP Deepwater Horizon oil spill. His research activities generally seek to understand community capacity to respond to and recover from disasters, as well as ways to enhance community preparedness and resilience.

Kathryn Hamilton is a graduate student in the Department of Sociology at Virginia Polytechnic Institute and State University. Her research interests include psychosocial impacts of hazards and disasters.

References

Aldhous, Peter; Lee, Stephanie M.; and Hirji, Zahra. 2021. "The Texas Winter Storm and Power Outages Killed Hundreds More People Than the State Says." Retrieved May 30, 2022, from https://www.buzzfeednews.com/article/peteraldhous/texas-winter-storm-power-outage-death-toll.

Arata, Catalina M.; Picou, J. Steven; Johnson, G. David; and McNally, T. Scott. 2000. "Coping With Technological Disaster: An Application of the Conservation of Resources Model to the *Exxon Valdez* Oil Spill." *Journal of Traumatic Stress* 13(1):23-39. https://doi.org/10.1023/A:1007764729337.

Benight, Charles C.; Ironson, Gail; Klebe, Kelli; Carver, Charles S.; Burnett, Kent; Greenwood, Debra; Baum, Andrew; and Schneiderman, Neil. 2007. "Conservation of Resources and Coping Self-efficacy Predicting Distress Following a Natural Disaster: A Causal Model Analysis Where the Environment Meets the Mind. *Anxiety, Stress, and Coping* 12(2):107-126.

Benight, Charles C.; Swift, Erika; Sanger, Jean; Smith, Anne; and Zeppelin, Dan. 1999. "Coping Self-Efficacy as a Mediator of Distress Following a Natural Disaster." *Journal of Applied Social Psychology*, 29(12), 2443–2464. https://doi.org/10.1111/j.1559-1816.1999.tb00120.x

Binder, Sherri B.; Ritchie, Liesel A.; Bender, Rose; Thiel, Alexis; and Baker, Charlene K. 2020. "Limbo: The unintended Consequences of Home Buyout Programs on Peripheral Communities." *Environmental Hazards* 19(5):488-507.

Bottner, Richard; Weems, John; Hill, Lucas G.; Ziebell, Christopher; Long, Sharon; Young, Sara; Sasser, Mike; Ferguson, Aaron; and Tirado, Carlos. 2021. "Addiction Treatment Networks Cannot Withstand Acute Crises: Lessons from 2021 Winter Storm Uri in Texas." *NAM Perspectives*. Retrieved May 30, 2022, at: https://www.ncbi.nlm.nih.gov/pmc/articles/PMC8406583/.

Campbell, Nnenia M.; Vickery, Jamie; and Ritchie, Liesel A. 2018. *Community Impacts and Risk Perceptions of Induced Seismicity, Report 1: Qualitative Findings*. Report prepared for the National Science Foundation, Award #1520846. Available from Liesel Ritchie at liesel14@vt.edu.

Clay, Lauren A. and Greer, Alex. 2019. "Association Between Long-term Stressors and Mental Health Distress Following the 2013 Moore Tornado: A Pilot Study." *Journal of Public Mental Health* 18(2): 124-134. doi:10.1108/JPMH-07-2018-0038

Davidson, Laura M. and Baum, Andrew. 1986. "Chronic Stress and Posttraumatic

Stress Disorders." *Journal of Consulting Clinical Psychology* 54(3):303-308.

Douglas, Erin and Ferman, Mitchell. 2021. "Another Texas power outage unlikely; This week, but severe weather this summer could prompt an electricity crisis." *The Texas Tribune.* Retrieved June 12, 2022, from https://www.texastribune.org/2021/04/15/texas-ercot-blackouts-summer-climate/.

Evensen, Darrick; Stedman, Richard; O'Hara, Sarah; Humphrey, Mathew; and Andersson-Hudson, Jessica. 2017. "Variation in Beliefs About 'Fracking' Between the UK and US." *Environmental Research Letters* 12(12) 124004.

Gill, Duane A., and Picou, J. Steven. 1998. "Technological Disaster and Chronic Community Stress." *Society and Natural Resources* 11:795-815.

Gill, Duane A.; Picou, J. Steven; and Ritchie, Liesel A. 2012. "The *Exxon Valdez* and BP Oil Spills: A Comparison of Initial Social and Psychological Impacts." *American Behavioral Scientist* 56(1):3-23.

Gill, Duane A.; Ritchie, Liesel A.; and Picou, J. Steven. 2016. "Sociocultural and Psychosocial Impacts of the *Exxon Valdez* Oil Spill: Twenty-Four Years of Research in Cordova, AK." *The Extractive Industries and Society* 3:1105-1116.

Gill, Duane A. and Ritchie, Liesel A. 2018. "Contributions of Technological and Natech Disaster Research to the Social Science Disaster Paradigm." In Rodriguez, Havidán., Donner, William., and Trainor, Joesph E. (Eds.), *Handbook of Disaster Research* (pp. 39-60). Springer, Cham. https://doi.org/10.1007/978-3-319-63254-4_3

Gill, Duane A.; Ritchie, Liesel A.; Picou, J. Steven; Langhinrichsen-Rohling, Jennifer; Long, Michael A.; and Shenesey, Jessica W. 2014. "The Exxon and BP Oil Spills: A Comparison of Psychosocial Impacts." *Natural Hazards* 74:1911-1932.

Glazer, Yael R.; Tremaine, Darrel M.; Banner, Jay L.; Cook, Margaret; Mace, Robert E.; Nielsen-Gammon; John, Grubert, Emily et al. 2021. "Winter Storm Uri: A Test of Texas' Water Infrastructure and Water Resource Resilience to Extreme Winter Weather Events." *Journal of Extreme Events* 2150022.

Green, Bonnie L.; Grace, Mary C.; Vary, Marshall G.; Kramer, Teresa L.; Gleser, Goldine C.; and Leonard, Anthony C. 1994. "Children of Disaster in the Second Decade: A 17-year Follow-up of Buffalo Creek Survivors." *Journal of the American Academy of Child and Adolescent Psychiatry* 33:71-79.

Halbesleben, Jonathon R.B. 2010. "The Role of Exhaustion and Workarounds in

Predicting Occupational Injuries: A Crosslagged Panel Study of Health Care Professionals." *Journal of Occupational Health Psychology* :1-16.

Halbesleben, Jonathon R.B. and Bowler, W. Matthew. 2007. "Emotional Exhaustion and Job Performance: The Mediating Role of Motivation." *Journal of Applied Psychology* 92:93-106.

Halbesleben, Jonathon R.B.; Neveu, Jean Pierre.; Paustian-Underdahl, Samantha C.; and Westman, Mina. 2014. "Getting to the 'COR': Understanding the Role of Resources in Conservation of Resources Theory." *Journal of Management* 40(5):1334-1364. https://doi.org/10.1177/0149206314527130.

Hobby School of Public Affairs. 2021. *The Winter Storm of 2021*. Hobby School of Public Affairs, University of Houston. Retrieved May 30, 2022, from https://uh.edu/hobby/winter2021/storm.pdf.

Hobfoll, Stevan E. 1989. "Conservation of Resources: A New Attempt at Conceptualizing Stress." *American Psychologist* 44(3):513-524.

Hobfoll, Stevan E. 1991. "Traumatic Stress: A Theory Based on Rapid Loss of Resources." *Anxiety Research* 4:187-197.

Hobfoll, Stevan E. 2011a. "Conservation of Resources Caravans in Engaged Settings." *Journal of Occupational and Organizational Psychology* 84:116-122.

Hobfoll, Stevan E. 2011b. "Conservation of Resources theory: Its Implication for Stress, Health, and Resilience." In Folkman, Susan. (Ed.), *The Oxford handbook of stress, health, and coping* (pp. 127-147). Oxford, England: Oxford University Press.

Horowitz, Mardi J. 1976. *Stress Response Syndromes*. New York: Aronson.

Horowitz, Mardi J.; Wilner, Nancy; and Alvarez, William. 1979. "Impact of Event Scale: A Measure of Subjective Stress." *Psychosomatic Medicine* 41(3):209-218.

Li, Dongying; Zhang, Yue; Li, Xiaoyu; Meyer, Michelle; Bazan, Marissa; and Brown, Robert. "'I Didn't Know What to Expect or What to Do': Impacts of a Severe Winter Storm on Residents of Subsidized Housing in Texas." Retrieved May 30, 2022, from https://papers.ssrn.com/sol3/papers.cfm?abstract_id=4047541.

Li, Xiaoyu; Zhang, Yue; Li, Dongying; Xu Yangyang; and Brown, Robert D. 2022. "Ameliorating Cold Stress in a Hot Climate: Effect of Winter Storm Uri on Residents of Subsidized Housing Neighborhoods." *Building and Environment* 209. Retrieved May 30, 2022, from https://doi.org/10.1016/j.buildenv.2021.108646.

Mäkikangas, Anne; Bakker, Arnold B.; Aunola, Kaisa; and Demerouti, Evangelia. 2010. "Job Resources and Flow at Work: Modeling the Relationship Via Latent Growth Curve and Mixture Model Methodology." *Journal of Occupational and Organizational Psychology* 83:795-814.

Morehouse, Catherine. 2022. "Spiking temperatures could cause more blackouts this summer. They won't be the last." *POLITICO*. Retrieved June 12, 2022, from https://www.politico.com/news/2022/05/31/spiking-temperatures-could-cause-more-blackouts-this-summer-they-wont-be-the-last-00034858/.

National Institute of Standards and Technology. 2016. *Community Resilience Planning Guide*. Retrieved June 10, 2022, from https://www.nist.gov/community-resilience/planning-guide.

Nejat, Ali; Solitare, Laura; Pettitt, Edward; and Mohsenian-Rad, Hamed. 2022. "Equitable Community Resilience: The Case of Winter Storm Uri in Texas." Retrieved May 30, 2022, from https://arxiv.org/abs/2201.06652.

Popik, Thomas and Humphreys, Richard. 2021. "The 2021 Texas Blackouts: Causes, Consequences, and Cures." *Journal of Critical Infrastructure Policy* 2(1):47-73.

Proffer, Erica. 2022. "Here is why death totals from Winter Storm Uri may vary." Retrieved May 30, 2022, from https://www.kvue.com/article/weather/winter-storm/here-is-why-death-totals-from-winter-storm-uri-may-vary/269-f2bf277f-74d9-443b-ab2e-ff89f336f3ec#:~:text=The%20Texas%20Department%20of%20State,the%20way%20deaths%20were%20counted.andtext=The%20State's%20report%20shows%20how%20it%20measured%20the%20amount.

Ritchie, Liesel A. 2012. "Individual Stress, Collective Trauma, and Social Capital in the Wake of the *Exxon Valdez* Oil Spill." *Sociological Inquiry* 82(2):187-211. https://doi.org/10.1111/j.1475-682X.2012.00416.x.

Ritchie, Liesel A.; Gill, Duane A.; and Long, Michael A. 2018. "Mitigating Litigating: An Examination of Psychosocial Impacts of Compensation Processes Associated with the 2010 BP *Deepwater Horizon* Oil Spill." *Risk Analysis* 38(8):1656-1671.

Ritchie, Liesel A.; Gill, Duane A.; and Long, Michael A. 2020. "Factors Influencing Stress Response Avoidance Behaviors following Technological Disasters: A Case Study of the 2008 TVA Coal Ash Spill." *Environmental Hazards* 19(5):442-462.

Ritchie, Liesel A.; Little, Jani; and Campbell, Nnenia M. 2018. "Resource Loss and Psychosocial Stress in the Aftermath of the 2008 TVA Coal Ash Spill." *International Journal of Mass Emergencies and Disasters* 36(2):179-207.

Ritchie, Liesel A. and Long, Michael A. 2021. "Psychosocial Impacts of Post-Disaster Compensation Processes: Community-Wide Avoidance Behaviors." *Social Science and Medicine* 270:113640.

Sattler, David N.; Whippy, Albert; Graham, James M.; and Johnson, James. 2018. "A Psychological Model of Climate Change Adaptation: Influence of Resource Loss, Posttraumatic Growth, Norms, and Risk Perceptions Following Cyclone Winston in Fiji. In Filho, Leal W. (Ed.), *Climate Change Impacts and Adaptation Strategies for Coastal Communities* (pp. 427-443). Cham, Switzerland: Springer.

Smead, Richard G. 2021. "ERCOT—The Eyes of Texas (and the World) Are Upon You: What Can be Done to Avoid a February 2021 Repeat." *Climate and Energy* 37(10):14-18.

Stipes, Chris. 2021. "New report details impact of Winter Storm Uri on Texans." Retrieved May 30, 2022, from https://uh.edu/news-events/stories/2021/march-2021/03292021-hobby-winter-storm.php.

Texas Department of State Health Services. 2021. *February 2021 Winter Storm-Related Deaths–Texas*. Retrieved May 30, 2022, from https://www.dshs.texas.gov/news/updates/SMOC_FebWinterStorm_MortalitySurvReport_12-30-21.pdf

United Stated Census. 2020. Retrieved June 17, 2022, from https://www.census.gov/quickfacts/TX.

Vogel, Eve and Vogel, Steven K. 2021. "The Texas Power Failure: How One Market Model Discovered Its Natural Limits." *Promarket*. Retrieved June 10, 2022, from https://www.promarket.org/2021/03/25/texas-power-outage-market-ercot-failure/.

Atlas for a Warp Speed Future: Enhancing Usual Operating Modes of the U.S. Government

Amanda Arnold[1]

[1] School for the Future of Innovation, Arizona State University, aarnold@asu.edu

Abstract

Operation Warp Speed (OWS) delivered new and effective vaccines to the general public in just 9 months, exploding previously held ideas about the government's role in medical countermeasure (MCM) development as well as what is possible on the timescale of vaccine development. OWS has potential to become a map for action in future pandemic crises. This article examines federal modes of governance that emerged in response to the Covid-19 crisis, with special attention to how those modes differ from normal government operations. It is at the intersection of crisis modes of action and normal modes of operation that lessons emerge from OWS that may be worth applying in normal times – or not.

In "Rules for Operating at Warp Speed," I outlined how the leadership of OWS was able to accelerate operations under a suspension of the government's usual modes of operation (Arnold, 2020[1]). This included suspension of rules that normally govern transparent and robust federal contracting and relaxing standards for scientific consensus-building and expertise across government. This article draws from interviews completed in 2020 and 2021 with senior officials at the Department of Defense (DOD), Food and Drug Administration (FDA), and the White House in order to identify the key pandemic modes of action contributing to the success of OWS. It also discusses whether (and how) those modes of action might be adapted to enhance critical infrastructure preparedness in non-crisis times.

Pandemic Modes of Action

When confronting the uncertainty, death, and social disruption of the Covid-19 pandemic, the normal modes of government operation were set aside in order to make room for crisis modes of action. Three modes of action emerged: Speed, Scale, and Scope.

---

1 https://issues.org/rules-operation-warp-speed-arnold/

Speeding Contracting using Other Transactional Authority: Driving Vaccine Development

As of July 2021, the Department of Defense (DOD), Department of Health and Human Services (HHS), and the Department of Homeland Security (DHS) obligated $12.5 billion in response to the Covid-19 pandemic through flexible contracting mechanisms, including Other Transaction Authority (OTA). According to a Government Accountability Office (GAO) report, OTA was routinely used to allocate funds in Operation Warp Speed in the name of acceleration. The report found that extensive use of this contracting authority mechanism lacked sufficient transparency and oversight (GAO, 2021[2]). This is because the OTA mechanism sweeps away standard government procedures usually valued as part of contracting rulebooks including the Federal Acquisition Regulations (FAR[3]) and the Defense Federal Acquisition Regulation Supplement (DFARS[4]). The difference between OTA and the traditional procurement regulations are stark: "the proverbial guidebook for OTA is only 53 pages long—incredibly brief in comparison to the FAR, a whopping 1,988 pages, and the DFARS, which comes in at 1,338 pages." (Arnold, 2022[5]). Interviews completed in conjunction with my doctoral research with late Trump and early Biden Administration officials (2020-20210) corroborated both the predominant use of these types of contracting mechanisms during OWS and a lack of accountability associated with OTA.

Prior to its expansive use during OWS, OTA was viewed as a potential abrogation of important administrative mechanisms that support the principled allocation of federal funding (Ardizzone, 2020[6]). Significant implications emerging from the extensive use of OTA during the pandemic include questions about the legal protections afforded by Bayh-Dole Regulations[7] for technology transfer and commercialization. These legal protections are closely tied to normal modes of federal contracting. The lack of transparency in OTA contracting could have been used to block government use rights or march-in authority (Douglass, 2021[8]).

The routine use of OTA in non-crisis times may threaten the standards of government procedures meant to ensure fairness and accountability of federal funding (Audit, 2021[9]). Further analysis is needed to support enhancing crisis funding mechanisms having the same robust standards of transparency and evi-

---

2 https://www.gao.gov/assets/gao-21-501.pdf
3 https://www.acquisition.gov/sites/default/files/current/far/pdf/FAR.pdf
4 https://www.acquisition.gov/sites/default/files/current/dfars/pdf/DFARS.pdf
5 https://issues.org/rules-operation-warp-speed-arnold/
6 https://www.keionline.org/wp-content/uploads/KEI-Briefing-OTA-29june2020.pdf
7 https://grants.nih.gov/grants/bayh-dole.htm
8 https://cshe.berkeley.edu/sites/default/files/publications/rops.cshe.3.2021.douglass.fedresearchbayhdolecovid.2.23.2021_1.pdf
9 https://media.defense.gov/2021/Apr/23/2002626394/-1/-1/1/DODIG-2021-077.PDF

dentiary support required in normal times. Likewise, there is a need to develop principled, novel funding mechanisms for use in normal times that can flex to accommodate crisis speeds. One avenue in seeking such approaches may be the growing interest in applying industrial policy to government modes of investment (Bonvillian, 2021[10]).

Scaling Conditional Drug: Flooding the Market Using Emergency Use Authorization

The Covid pandemic tested FDA's accelerated emergency capacity on a massive scale, with FDA issuing conditional approval for over 400 tests, vaccines, and antiviral drugs in the first 13 months of the pandemic (Parasidis, 2021[11]). The FDA was able to scale to this approval frequency by utilizing a critical crisis legal authority called Emergency Use Authorization (EUA).[12] EUA may only be deployed following emergency declaration by the President or his appointees. In 40 days of February and March 2020 Secretary of Health and Human Services Alex Azar exercised this authority making three emergency declarations.[13] This authority allows FDA to approve promising countermeasures as they show promise earlier on and works by spreading risk in clinical trial design across pre-clinical and post-market authorization. The goal is getting products to patients who would otherwise die without a medical countermeasure (MCM) (FDA, 2022[14]).

EUA is a relatively new regulatory tool at FDA only codified in the Project Bioshield legislation of 2004.[15] The first EUA was approved for an Anthrax vaccine in 2005 (Federal Register, 2005[16]). Expanded as part of the Public Readiness and Emergency Preparedness Act (PREP Act[17]) of 2005, the EUA was used sparingly until the swine flu pandemic of 2009 when 22 EUAs were approved (Iwry, 2021[18]). Several pre-emptive EUAs were also issued for Ebola, Zika, and MERS, though no effective treatments or cures were identified (Bobrowski, 2020[19]). There is much work still to be done to study the challenges associated with this massive expansion of the EUA mechanism during the Covid-19 pandemic. For the purposes of this work, the EUA reflects an important pandemic mode of

---

10 https://itif.org/publications/2021/10/04/emerging-industrial-policy-approaches-united-states/
11 https://www.fdli.org/2021/12/assessing-covid-19-emergency-use-authorizations/
12 https://www.law.cornell.edu/uscode/text/21/360bbb-3
13 https://blog.petrieflom.law.harvard.edu/2021/01/28/fda-emergency-use-authorization-history/
14 https://www.fda.gov/media/142749/download
15 https://www.govinfo.gov/content/pkg/PLAW-108publ276/pdf/PLAW-108publ276.pdf
16 https://www.federalregister.gov/documents/2005/02/02/05-2028/authorization-of-emergency-use-of-anthrax-vaccine-adsorbed-for-prevention-of-inhalation-anthrax-by
17 https://aspr.hhs.gov/legal/PREPact/Pages/default.aspx
18 https://www.fdli.org/2021/09/fda-emergency-use-authorization-a-brief-history-from-9-11-to-covid-19/
19 https://www.ncbi.nlm.nih.gov/pmc/articles/PMC7361119/

action in which scaling the normal federal approval process required additional authority.

The EUA mechanism expires with the emergency declaration(s) that authorized its use. Normal modes of operation within the FDA allow for at least four non-crisis mechanisms designed to accelerate the approval of drugs and vaccines including accelerated approval for serious conditions and expedited development.[20] While these non-crisis mechanisms cannot meet the scale of new candidates explored during the Covid pandemic, more assessment is essential to enhance the EUA mechanism. For instance, the EUA path may allow pressure by influential political leaders on conditional approval of drugs widely seen as ineffective or even dangerous. This was the case in the FDA's EUA approval of hydroxychloroquine and chloroquine for conditional use in hospitals in late May, 2020. The approval was revoked in June (FDA, 2020[21]). Despite the comparatively quick revocation of the approval, the close connection between FDA's EUA issued for the application of these malaria drugs to Covid-19 – and the President's statements on their supposed effectiveness – damaged the reputation of the FDA approval process (Science 2020[22]). This concern for the political pressure on FDA via the use of EUA was corroborated in my own interviews with senior OWS leadership.

Expanding Scope to Product Development: Beyond the Linear Model of Federal Investment

The scope of OWS expanded federal funding infrastructure beyond the normal modes of operation. Funding was pushed far in the direction of product development and steps done in parallel rather than the usual process of waiting for a prototype, then lead product, and then progressing stepwise through clinical trials. In non-crisis, according to this linear model of innovation that has governed federal R&D since WWII, the federal government normally invests heavily in discovery science and pre-clinical development of medical countermeasures through mechanisms such as R01 (investigator-initiated) grants at the National Institutes of Health. Government typically provides less support for subsequent steps in development and marketing, leaving those steps to small company formation, technology transfer between universities and industry, and R&D investment in industry to further develop and commercialize research leads into actual products.

This point is especially important in relation to OWS, which did not facilitate the *invention* of a vaccine to curb Covid but rather *developed* existing

---

20 https://www.fda.gov/patients/learn-about-drug-and-device-approvals/fast-track-breakthrough-therapy-accelerated-approval-priority-review

21 https://www.fda.gov/drugs/drug-safety-and-availability/fda-cautions-against-use-hydroxychloroquine-or-chloroquine-covid-19-outside-hospital-setting-or

22 https://www.science.org/content/article/former-fda-leaders-decry-emergency-authorization-malaria-drugs-coronavirus

candidates. This point is corroborated by OWS leaders Moncef Slaoui and Matt Hepburn who wrote that the strategy for OWS was to select existing vaccine candidates held by industry that used one of four vaccine-platforms including mRNA; replication-defective live-vector; recombinant-subunit-adjuvanted protein; or attenuated replicating live-vector. Many of the efforts to make a SARS-CoV-2 vaccine emerged from moving selected candidates through phase 2-3 clinical trials, approval, and commercialization (Slaoui, 2020[23]).

The government made this investment in Covid treatments and vaccines through an expanded scope of federal investment not seen since WWII. The massive financial cost of OWS was $18 billion in just over one year, an expenditure on par with the Manhattan Project, which built the atomic bomb at a cost of $23 billion over 5 years (inflation-adjusted) (Shulkin, 2021[24]). Similar to the Manhattan Project, OWS was a development effort, not a research project.

It is clear from the experience in OWS that government investment in this final stage of development, where the science is developed over the decades preceding, does speed the movement of new vaccines and other medical countermeasures from industry labs to patients awaiting much-needed medical interventions. Given the likelihood of pandemic crisis-non-crisis oscillation, extending the scope of federal investment into the final stages of development should move more products from the lab to the market, providing more value to patients.

Adapting Pandemic Modes of Action to Critical Pandemic Preparedness Infrastructure

The National Academies of Science recently released a report on aspects of the government-wide response to the pandemic stating, "[medical countermeasure] preparedness and response requires an enterprise that manages resources efficiently in day-to-day work, without compromising on quality." (NASEM, 2021[25]) The key to enhancing medical countermeasure (MCM) development in the U.S. government is through enhancing robust, transparent, and elastic mechanisms that function in both crisis and non-crisis at the necessary speed, on the necessary scale, and with necessary scope to develop the medical products that are needed.

Accountable and transparent funding infrastructures for product development are needed that are sufficiently elastic to support the speed and flexibility required during crisis

Other Transaction Authority (OTA) will likely continue as an elastic contracting

---

23 https://pubmed.ncbi.nlm.nih.gov/32846056/
24 https://catalyst.nejm.org/doi/full/10.1056/CAT.21.0001
25 https://nap.nationalacademies.org/catalog/26373/ensuring-an-effective-public-health-emergency-medical-countermeasures-enterprise

mechanism to expand medical product development funding. However, while OTA proved essential for rapid test, drug, and vaccine development during the pandemic, it also subverts important principles underlying normal contracting procedures. In the short term, the key lever should not be sole reliance on after-action reporting to ensure transparency and ethical spending. A data-based approach to capturing in real time who is being funded and under what reasoning and by whom – by way of a dynamic crisis dashboard – is critical. Such an analysis should be transmitted to the Office of the Assistant Secretary for Preparedness and Response (ASPR) as well as the Office of Management and Budget (OMB) at regular intervals during crisis. The dashboard should also be made available to the public.

The accountability of federal agencies, including the FDA, cannot be sacrificed during crisis response as scale

The EUA is an authority that enabled a scale of approvals to meet the pandemic need that would not have been otherwise been possible. However, the emergency declaration that triggered this new approval authority by FDA also contributed to delay. This is because the CDC's first approved test for Covid experienced an issue with the reagent and no other test had been created nor approved by FDA. The emergency declaration required emergency approval by FDA whereas this emergency approval by FDA would not have been required prior to the emergency declarations.

The importance of diagnostic testing at the outset of the pandemic cannot be overstated. The pandemic declarations, and the FDA authorities that ensued, did also create a bureaucratic hurdle that significantly slowed early response (Science, 2020[26]). In addition, and as outlined above, the EUA authority itself was used as a political tool by the President when FDA allowed a controversial drug, hydroxychloroquine, to be used as a therapeutic, leaving a deficit of accountability in its wake. The testing issue must be addressed for the future. The question of how to prevent the politicization of the EUA authority in future crises must also be considered.

Federal Funding for medical product development is an untapped opportunity to speed medical countermeasures to patients

The current model of development for medical countermeasures, especially related to emerging and infectious disease, will not be sufficient to meet future pandemic preparedness and response needs (Vu, 2022[27]). There is opportunity for new approaches that leverage government investment and endorsement to actually create

---

26 https://www.science.org/content/article/united-states-badly-bungled-coronavirus-testing-things-may-soon-improve

27 https://alomit.wpengine.com/wp-content/uploads/2020/04/P0695-1.pdf

and increase value in markets that otherwise may not be attractive to industry (Laplane, 2020[28]). The experience during OWS suggests that the traditional model of federal funding for basic and early applied research, depending on private capital for late-stage development, can rapidly meet non-crisis health needs if scope of federal funding support is expanded all the way through development with serial process collapsed into parallel processes along the way.

There is already a suggestion for how to fund this expanded scope of federal research and development infrastructure. Using the principles of financial engineering and securitization, Andrew Lo of MIT suggests the development of a fully leveraged megafund to organize and grow support across a series of medical candidates. This approach would mitigate the risk of failed investments by the government by leveraging the likelihood of successful investments. If the fund is large enough and based on models of the megafund completed to date, the returns could yield a profit of up to 8 percent for the government and industry investors (Fangnan, Yang, and Lo, 2015[29], 2013[30] and Lo, 2021[31]). An additional benefit of having a concerted government effort to expand government R&D would be the opportunity to establish evaluation practices at the outset to measure the success of such efforts through evidence-based policy (Baron, 2018[32]).

Conclusions

Normal modes of government operation associated with accountability and transparency were relaxed during the Covid pandemic crisis to allow new modes of action associated with speed, scale, and scope to emerge. As the pandemic threat continues, policy actions are needed to bring these two extremes into harmony. Several of the policy recommendations discussed here – including accountable crisis contracting mechanisms; the maintenance of principled federal agency actions; and the expansion of federal government in support of product development – would enhance the harmony between normal and crisis modes. The study of Operation Warp Speed, including what worked and what did not work, provides an important atlas to navigate a future of crisis/non-crisis oscillation in a way that will be less disruptive and more manageable than the crisis approach we just experienced.

---

28 https://www.sciencedirect.com/science/article/pii/S2590145120300025#bib0220
29 https://www.science.org/doi/abs/10.1126/scitranslmed.aaa2360
30 https://www.aeaweb.org/articles?id=10.1257/aer.103.3.406
31 https://jsf.pm-research.com/content/27/1/17.abstract
32 https://journals.sagepub.com/doi/abs/10.1177/0002716218763128

Author Capsule Bio

Amanda Arnold is a policy practitioner in Washington DC who has worked in the federal government, academia, and industry. This includes work in the vaccine development sector for over 15 years. A published author, she holds a master's degree in Science and Technology Policy and is a Doctoral Candidate with graduation anticipated in 2022.

Strengthening the Security of Operational Technology: Understanding Contemporary Bill of Materials

Arushi Arora,[1,2] Virginia Wright,[3] Christina Garman[4]

[1] National and Homeland Security Laboratory, Idaho National Laboratory
[2] Corresponding Author, Arushi.Arora@inl.gov
[3] Energy Cybersecurity Portfolio Manager, Idaho National Laboratory
[4] Assistant Professor, Department of Computer Science, Purdue University

Abstract

The evolution of cyber-physical infrastructure has made its security more challenging. The last few years have witnessed a convergence of hardware and software segments in various domains, including operational technology (OT) which is responsible for carrying out critical tasks such as monitoring and controlling power grids, nuclear plants, transportation, and emergency services. Both hardware and software encapsulate numerous open source and proprietary subcomponents, making it crucial for end-users to understand the composition of the products they are using. For example, wind turbines incorporate thousands of lines of code (software) used for the turbine's design, planning, operation, and analytics in addition to the numerous hardware subcomponents that construct it. Due to the highly complex nature of software and hardware, knowledge of the components and subcomponents is required to mitigate cyber vulnerabilities and to defend against cyberattacks.

There has also been a transformation from a traditional linear supply chain into a global, dynamic, diverse, and interconnected system. The digitization of the supply chain makes it easier to find and exploit vulnerabilities. Critical infrastructures (e.g., power grids, oil, natural gas, water, and wastewater) rely on OT to function, and if the OT is compromised, equipment damage and potential interruption of services could result. A significant security measure to protect OT systems from disruption is to develop a supply chain bill of materials (BoM) corresponding to the software and hardware used in OT, along with attestations amongst vendors and asset owners. A supply chain BoM is a proactive way to understand the inherent vulnerabilities in the system and mitigate them in advance of being exploited. BoMs bolster the trust placed in the digital infrastructure and enhance software supply chain security by sustaining the management of component obsolescence and compliance, along with the seclusion of unsafe segments of a specific product.

Adopting BoM tools is becoming increasingly important across various government sectors, as evidenced by the recent U.S. executive order on cybersecurity (NIST 2021). This paper aims to classify BoMs based on structure, functionality, component type, and architecture. The work also discusses case studies to further highlight the benefits of BoMs. In addition, it identifies missing pieces in existing BoM implementations so that future research may identify bounds on where it could expect to make improvements and directly enable researchers to identify promising areas for exploration. Further, the authors provide valuable recommendations to tool developers, researchers, and standardizing organizations (policymakers), additionally benefitting critical infrastructure owners and government executives. This aids in paving a path for future work, thereby, providing suggestions to determine a tool for consumers that best suit their needs.

Keywords: Software Bill of Materials (SBoM), SBoM tools, supply chain, operational technology, critical infrastructure security

Introduction

The 20th century brought the Digital Revolution, which led to the adoption and proliferation of digital computers and digital record-keeping. The past few years have also witnessed the introduction of online shopping, connected vehicles, smart homes, and technologies that have completely disrupted the conventional monolithic supply chain. This has also inflated consumers' demands and expectations, for instance in the form of fast delivery. The escalating infiltration of digital technologies such as artificial intelligence (AI), blockchain, and automation has also generated vast opportunities for organizations and supply chain practices.

Traditional supply chains, which were meant to be linear, have transformed into a dynamic, diverse and interconnected system. Most major organizations are now scattered so far and wide that they have supply chains that exist in multiple countries. These vast supply chain operations can range from customer service obligations to substantial critical infrastructures; therefore, the circumstances disrupting them can prove catastrophic.

A bill of materials (BoM), which defines the components that are needed to manufacture a product, can help prevent bottlenecking the supply chains, ensure faster time to market, guarantee greater operational efficiency and may lessen the risk of errors and rework. BoMs can better represent the supply chain and associated documents and payloads contained at each stage. The contemporary digitalized BoMs seek to improve supply chain efficiency by facilitating sharing of BoM-relevant information, including data licenses, software descriptions, attestations and versions, hardware components, and lists of staff or other human resources, among supply chain partners.

BoMs support the management of component obsolescence and compliance and the isolation of unsafe segments of a specific product, strengthening the trust placed in the digital infrastructure and enhancing software supply chain security. It also provides an autonomous way to apply government regulations with contractors to protect against the use of counterfeit devices and compromised software in critical infrastructure. To operate correctly, critical infrastructure further relies on operational technology (OT) solutions, which aim to regulate industrial equipment, building management systems, fire control systems, and physical access control mechanisms. Implementing a mechanism to produce, consume, and transport BoMs is especially important to safely monitor OT components that are used in multiple areas like nuclear power plants, wind farms, and power grids, (Parnas, Asmis, and Madey 1991; Slootweg et al. 2003; Shiroudi et al. 2012; Singhal and Saxena 2012).

OT components are usually not replaced at the same rate as consumer technology infrastructure, and therefore lack modern cyberattack preventive measures. This raises concern about cybersecurity vulnerabilities existing in components or devices supporting critical infrastructure, as cyberattacks can cause equipment damage, interruption of services, and even loss of life. It is crucial to have a mechanism to share BoMs, corresponding to the software and hardware used in respective components, along with attestations amongst vendors and asset owners. BoMs can ensure all parties involved in the digital supply chain network—including suppliers, manufacturers, and customers—are aware of changes concerning their product of interest through automated synchronization.

In this paper, the authors classify BoMs based on structure, functionality, component type, and architecture. They further provide a deep insight into BoMs that will help readers identify tools that best suit their requirements based on the tool's utility. This paper also specifies broad use cases of employing BoMs and highlights the benefits. In addition, the authors identify missing pieces in existing BoM implementations so that future research may identify bounds on possible improvements and directly enable researchers to identify promising areas for future exploration. The authors also argue that this domain still lacks research thereby paving a path forward for future work.

Overview

Adopting SBoM tools is becoming increasingly important across various government sectors, as evidenced by the recent U.S. executive order on cybersecurity, EO 14028 (NIST 2021). In 2018, the National Telecommunication and Information Administration (NTIA) gathered a cross-sector, industry-led, multi-stakeholder process on software component transparency, to understand the potential, obligations, and obstacles in the domain (NTIA 2021d). Since its initiation, experts have made valuable progress in defining an SBoM and listing its benefits, identifying existing standards and formats that can be employed to convey information (NTIA 2021e; NTIA 2021g). This section provides a brief overview of the BoM, highlighting its motivation, the process involved, and its use cases, along with an example.

BoM

A BoM is a list of ingredients that make up a particular product. An SBoM identifies and lists software components, whereas a hardware BoM (HBoM) identifies and lists hardware components of a product. With the convergence of software and hardware in modern devices and instruments, their respective BoMs must identify and list both software and hardware components that constitute the device. The amount and type of data included in each BoM may differ and may depend on factors such as the BoM's usage, its industry or sector, and the needs of BoM consumers. NTIA (NTIA 2021a) gives a brief overview of actors in an SBoM system and defines a set of baseline SBoM components as listed in Table 1 and 2, respectively.

Motivation

Modern software often reuses and imports open-source components, such as libraries, from a previous stage, combining it with novel contributions to build a different product. The present digital supply chain is an intricate web of dependencies and any modification in the chain can result in wide-ranging consequences. It is hard to evaluate if the software product meets consumers' standards, obeys licensing regulations, or is free of known vulnerabilities. For example, in addition to hardware components, wind turbines incorporate software that may include third-party or proprietary software as subcomponents, which may be responsible for the turbine's design, planning, operation, and analytics (shown in figure 1). With such a complex and diverse nature of contemporary supply chains, it is hard to mitigate a disruption that occurs somewhere in the chain. Another challenge is to upgrade downstream component(s), in case of a patch in its corresponding upstream element.

Process

The first step to creating an SBoM during the software build and packaging process is to record the components that the supplier creates themselves along with its related information and all the reused imported components. The SBoM provides additional information about the software's subcomponents' technical variations, versions, and underlying inherited vulnerabilities. An SBoM is created whenever a component undergoes updates, upgrades, releases, and patches. Next, the SBoM must be made available to the consumer along with the software component. This may assist the consumer in performing an impact and compliance assessment along with the software's validation. The retrieval of the SBoM should be simple and can be carried out by providing an additional file, or a dynamic URL which may be beneficial for constrained devices. This facilitates a persistent assessment of the software and tracking of any updates.

Figure 1. Hardware and software integration in a wind turbine.

Use Cases

From the perspective of a supplier, an SBoM can help monitor the software's components for vulnerabilities, their end-of-life (EOL), and understand their complex dependencies, thereby providing better transparency into the code resulting in prompt delivery of code updates or patches (NTIA 2021e). It can also support the following: aid in reducing code bloat by providing complete knowledge about relevant and available open source components; familiarize the supplier with the license obligations of the components that are used in a software product; and, aid suppliers in reviewing code and identifying blacklisted components.

From the perspective of a consumer, an SBoM can support verification of the software components' sources, check for compliance with the consumer's policies, and scan for known vulnerabilities by performing software component analysis. It can also help the consumer to identify requirements for the software product as well as its integration. Further, an SBoM can provide a consumer with additional data, for instance, the software components' EOL, and attest claims that the supplier makes. An SBoM may also help in stimulating independent mitigations and make versed risk-based decisions while operating the software. It also keeps information more organized and readily available, which may aid in effective planning and reducing overall costs through a more contoured and efficient execution.

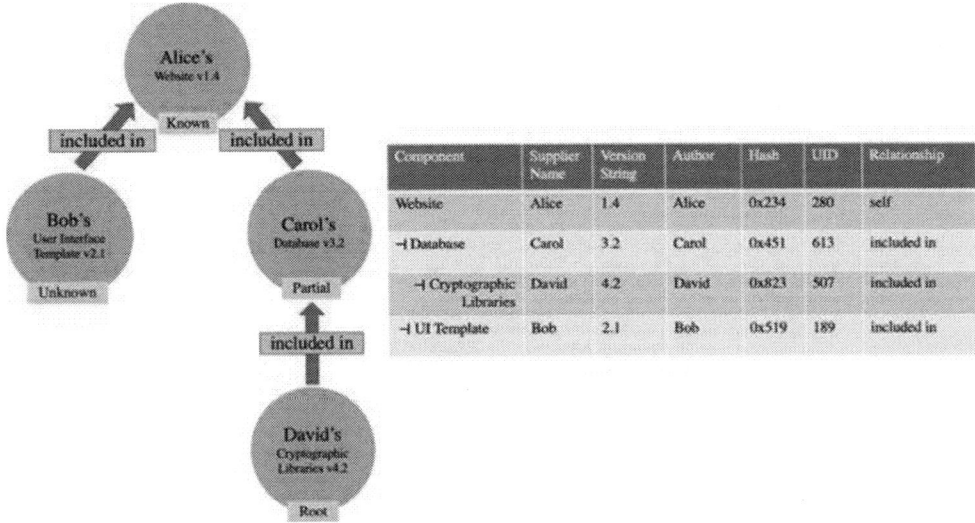

Figure 2. Conceptual SBoM tree and table with upstream relationship assertions.

An Example

Figure 2 presents a conceptual SBoM dependency tree along with a table stating all SBoM elements (as described in Table 2) that belong to Alice's application. As per the SBoM information, this application has four components: Alice's application (*primary*), which reuses Bob's user interface template, Carol's database application, and David's cryptographic libraries. The upstream component (Bob's user interface template) has unknown dependencies. In other words, Bob's code may be the root or if it may have further upstream components. Alice's website also uses Carol's database application which further employs David's cryptographic libraries. In addition, it is known that David's library is the root and does not utilize any upstream components.

Terminology

Various organizations have launched both open source and proprietary BoM tools. Previous works in this domain do not thoroughly investigate and describe existing BoM terminologies, their classifications and formats, and desired properties in the tools. In this section, the authors attempt to elaborate on this aspect of the contemporary BoM.

Table 1. SBoM Actors (NTIA 2021d).

Actors	Description
Element	Ingredient of an SBoM system
Supplier	The entity of the SBoM that creates, defines, and identifies ingredients of a software product and generates associated SBoMs
Component	The unit of a software product. A component may be a library, a file, or larger software like an operating system, a database system, or an office suite that is further comprised of components. It is defined by a supplier when the component is created, packaged, or distributed.
Consumer	An entity that receives SBoMs
Operator	An operator is a leaf entity in the SBoM conceptual tree that is a consumer but not a supplier. (Note that a consumer who reuses components can also be a supplier.)
Author	An author makes claims about components created or incorporated by a separate entity i.e., a supplier
Attribute	Properties and knowledge about a component
SBoM Entry	A row in the SBoM table specifying a component and relevant attribute

Classification

In recent years, hardware and software have converged and are dependent on one another such that the two are no longer mutually exclusive. In addition to classifying BoMs based on their structure and functionality (Zhou et al. 2018; Liu, Lai, and Shen 2014), the authors introduce a classification category based on their component type. BoMs can contain sensitive information which may not be public and require restricted access, so the authors also introduce another classification category to accommodate this criterion based on a BoM's availability.

Based on Structure

This type of BoM is suitable when the product does not constitute subcomponents. A single-level BoM is utilized for non-complex products with just one level of subassemblies. On the other hand, a multi-level BoM is used in more complicated assemblies of a final product and is, therefore, composed of multiple levels of subcomponents.

Based on Functionality

Functionality BoMs are broadly classified into eight categories.

Table 2. List of baseline SBoM Elements (NTIA 2021d).

SBoM Element	Description
Author Name	Author of the SBoM entry
Supplier Name	Name or identity of supplier of the component in the SBoM entry
Component Name	One or more component name(s). May include the capability in case of multiple names or aliases
Version String	Version information which aids in identifying a component
Component Hash	Cryptographic hash of the component to uniquely and precisely identify a binary
Unique Identifier	A unique identifier of the component
Relationship	Relationship is inherent in the design of the SBoM. This could be *includes* (downstream) or *included in* (upstream). A *downstream* component is the one towards the consumer whereas an *upstream* component is towards the supplier.
Relationship Assertions	Refers to a component representing its immediate upstream relationships. Four categories cover these assertions— • *Unknown*: This implies that immediate upstream (towards the supplier) components are not currently known and therefore not yet recorded. • *Root*: This indicates that there are no immediate upstream relationships. • *Partial*: There is at least one immediate upstream relationship and others may or may not be known. Known relationships are documented. • *Known*: The entire set of immediate upstream relationships are known and documented.

1. Configurable BoMs contain a list of all the components required to devise a configurable product, to meet customers' particular requirements.

2. Manufacturing BoMs provide a roadmap for manufacturing any physical product, presenting a structured list of all the components and subcomponents, including their relationships with one another.

3. Engineering BoMs provide the necessary elements and directions for making a particular product from a design standpoint and further helps in composing the manufacturing BoM.

4. Sales BoMs list the final finished product as a sales item and its respective subcomponents as inventory.

5. Assembly BoMs are similar to the sales BoM as they contain the finished product mentioned in the sales document. However, this document does not list the subcomponents of the final product.

6. Service BoMs carry a list of all the components, installation steps, and restoration directions. This BoM is used by service technicians while installing or servicing a product.

7. Production BoMs list subcomponents, costs, specifications, and quantities of a finished product, serving as the groundwork for a production order. The components listed can also be converted into finished items during the production process.

8. Template BoMs allow flexibility to update quantities or replace components of the product and can be used for either production or sales BoMs.

Based on Component Type

SBoM is effectively a nested inventory, and a list of components that make up the software. Similarly, an HBoM provides a nested list of all the elements that make up the hardware.

Based on Availability

A public BoM architecture is completely open and can be accessed by anyone. A private BoM architecture is accessible to only selected members of a group or an organization, thereby serving a private business. A permissioned BoM architecture, which lies between the above-listed categories, allows a verified entity to access the BoM, thereby accommodating different organizations sharing a supply chain.

Storage Mechanism

Finding a suitable database solution for any application is not straightforward, and this challenge holds true for BoMs. Application developers choose between relational, non-relational, and blockchain-based databases based on the type of data handled by the application. Other factors to consider include operation and maintenance costs, service stability, performance, scalability, security, and ease of developing database interface and modifying the database schema (Jatana et al. 2012; Boicea, Radulescu, and Agapin 2012; Aboutorabi et al. 2015; Li and Manoharan 2013).

Properties

Table 3 gives an overview of the SBoM tool taxonomy, which is significant to the range of use cases in SBoM generation and consumption. Tool consumers may benefit from identifying and understanding availability as per their needs. The information in this table has been derived from categories defined by NTIA (NTIA 2021c) in addition to the *transport* category, which represents the tools that support a mechanism to share and exchange SBoMs (NTIA 2021b). For instance, an organization that develops and vends software might be interested in SBoM tools that fall under the produce category, whereas the consumers of this software might be interested in SBoM tools that fall under the consume category. Further, the authors provide a list of security and privacy properties in Table 4 that may be desired by the users to support SBoMs containing sensitive information. This may be beneficial for tool developers and consumers while developing and choosing a tool for use.

Table 3. Taxonomy of a BoM tool (NTIA 2021c).

Category	Type	Description
Produce	Build	Automatic BoM creation as part of building a software artifact containing information about the build
	Analyze	Analysis of source or binary files generates the BoM by inspection of the artifacts & associated sources
	Edit	Assists manual entry or allows BoM data editing
Consume	Diff	Enables comparison of multiple BoMs
	View	Allows understanding of the contents in human-readable form, supporting decision making
	Import	Supports discovering, retrieving, and importing a BoM into the system for further processing and analysis
Transform	Translate	Permits file type conversion while preserving the same information
	Merge	Allows combining multiple sources of BoMs and other data together for analysis and audit purposes
	Support	Support use in other tools by application programming interface (APIs), object models, libraries, transport, or other reference sources
Transport	Advertise	Specifies how software or a device informs consumers that an SBoM is available
	Discover	Describes how the consumer learns of the location of the SBoM
	Access	Enables the transfer of the SBoM using the method derived from discovery

Table 4. Desired privacy and security properties in a BoM tool.

Category	Property
Encryption	Capability to encrypt data at rest, which is not being actively used and is stored in one location on hard drives, laptops, flash drives, or cloud storage
	Ability to encrypt data in transit, which is data active and flowing between devices and networks
Authentication	Ability to authenticate individual users and groups
Authorization	Ability to authorize individual users and at organizational level
	Capability to enable role-based access
Accounting	Ability to identify an executed operation and the *user* and its *organization* who executed the respective operation
	Ability to timestamp activities
	Ability to access client metadata
Auditability	Ability to provide transaction-level logging
	Capability to ensure tightly controlled deletions

State-of-the-Art

Many organizations rely on open-source software for their product development. A conventional software product nowadays often reuses third-party libraries and packages as subcomponents, which are not obvious to detect. SBoM tools can aid in understanding dependencies within more comprehensive and complex software, comply with the licenses, review software's components for vulnerabilities, blacklist components if required, and provide SBoM to consumers. To accomplish this goal, SBoM standards, such as Software Package Data Exchange (SPDX) (Gandhi, Germonprez, and Link 2018), Software Identification (SWID) tags (Waltermire and Cheikes 2015; SWID Tags, n.d.) and CycloneDX (CycloneDX, n.d.) have been introduced (NTIA 2021c).

Why Use SBoMs?

SBoMs can prove to be beneficial in multiple industries including automotive, software, agricultural, healthcare, and energy sectors. They can decrease unplanned, unscheduled work, in addition to reducing code bloat. Below are some of the benefits that SBoMs may provide.

Vulnerability Management

An SBoM may be beneficial to determine if a subcomponent contains any vulnerabilities and needs a patch. This also helps identify any downstream component that may be at risk due to this vulnerability. Further, an SBoM may provide details of previous patches and updates. Introducing and automating vulnerability assessment and management can curtail time to remediation and promote awareness (Dwyer 2018; Greenberg 2018).

License Management

SBoMs can guarantee that the software and its respective subcomponents comply with the licensing requirements.

Intellectual Property

SBoM information can assist intellectual property applications, as access to such data provides details of subcomponents used, licenses, and history of updates. It is worth noting that both SPDX and SWID can provide license information.

High Assurance

SBoMs can ensure the integrity of subcomponents in the software by providing the source, information about its suppliers, history of modifications, and the chain of custody as components move through the supply chain. It also stimulates the exclusion of low-quality or abandoned software elements. Further, the SBoM also includes information about the technologies supported by the corresponding software and EOL dates which may assist in carrying out timely updates to the software.

To help readers understand how this research can benefit in averting severe cyberattacks that affect critical infrastructure, the authors shed light on the recent attack on SolarWinds and Log4j (CISA 2021) vulnerability along with a discussion on their prevention (Wolf, Growley, and Gruden 2021).

Case Study: The Attack on SolarWinds

The Attack

SolarWinds is an American company that provides software for monitoring network and information technology (IT) infrastructure, serving hundreds of thousands of organizations around the globe. In September 2019, one of SolarWinds' software products, Orion, was infiltrated with malicious code, known as SUNBURST, by hackers who discreetly broke into the company's systems. This malware, which remained hidden when dormant, could create backdoors when active, permitting a malicious user to gain unauthorized access to the software ecosystem. This further let the hackers gain access to the software's host machine and other components in the network. The infected Orion software was available as a legitimate SolarWinds software update which went undetected, impacting about 18,000 SolarWinds' customers, both government and commercial. SolarWinds' supply chain attack has also affected contractors that do not use the company's software. For instance, CrowdStrike and Malwarebytes, which are not SolarWinds' customers, were indirectly affected through third parties.

Prevention

While the usage of SBoMs was not widespread and their benefits were not widely understood at the time (as Executive Order 14028, NIST 2021 was partly in response to such cyberattacks, CSIS 2022), the authors argue that such damaging supply chain cyberattacks could potentially be thwarted in the future by enforcing certain preventive measures, including the implementation of SBoMs. If a vendor, like SolarWinds, had provided an SBoM along with the software (or its update) to its customers, the latter could have performed various inspections that might have prevented this attack. An SBoM could assist customers in verifying the software's integrity. For instance, the hash of the software component in the SBoM can verify that the respective component has not been tampered with. The SBoM can also list hashes computed from multiple algorithms. Therefore, the customer could have further verified the software's authenticity. SBoMs can provide the vendor's digital signature, enabling the customers with non-repudiation capabilities. In addition to allowing the consumer to understand the components and third-party subcomponents of the software, SBoMs can also verify license compliance and check known vulnerabilities existing in the respective software.

Case Study: The Log4j Vulnerability

The Attack

This case study discusses the remote code execution vulnerability in Apache Log4j (a logging library for Java) disclosed on December 10, 2021 (CISA 2021). This library is a subcomponent used by many companies in a variety of Java-based software and web portals. The vulnerability allowed an attacker to execute arbitrary code on devices or execute a DoS (Denial of Service) attack on the targeted servers, remotely. This allows a malicious entity to further steal the data or take control over the affected target system.

Prevention

This case study is a perfect instance to showcase how modern software is composed of multiple third-party and open-source elements making it complex (Bauer et al. 2020). The widespread use of Java across both IT and OT platforms made this vulnerability complicated to patch as many open-source, as well as proprietary applications, also utilized the Log4j library. In addition, Log4j is also used as a JAR file in a third-party application and is also embedded in many custom applications. Therefore, remediating such an attack is complicated along with being resource-intensive.

It is, therefore, feasible to implement measures in order to safeguard against such vulnerabilities beforehand. The authors argue that keeping a record of all components enables organizations to analyze the impact, identify, and mitigate risks allowing greater transparency. The authors claim that mandating an SBoM along with the software purchase could have helped many organizations deal with such a vulnerability promptly, aiding them to perform a targeted search by scanning just the SBoM. Having an SBoM along with the software would have also made the companies more aware of the subcomponents being reused in the respective software.

Building the Right SBoM

This paper's systemization and categorization of SBoM tools, and their features could have helped SolarWinds (the vendor) to build an SBoM to achieve the above. First, observe a few key properties (Produce, Consume, or Transform) and features (SBoM format, repository support, user interface, and platform support) the vendor might want, based on the discussion in the previous section. The vendor could have chosen a produce software, which at least sustains the build feature supporting automatic BoM creation as part of building a software artifact. Additionally, the vendor could combine the build property with consume properties (if desired) which can support understanding, comparing, discovering, retrieving, and importing SBoMs. Transform properties could have been combined which could support translating SBoMs from one file type to another, merging multiple SBoMs, and assisting use in other tools through APIs. Further, the vendor could have considered other suitable features like format (CycloneDX), user interface (CLI), cross-platform, and API support. Verification and analysis tasks on the consumer side could have been achieved by choosing SBoM tools that support the Analyze (Produce) property, which generates the BoM by inspection of the artifacts and associated sources, and the View (Consume) property, which allows understanding of the contents in human-readable form, endowing decision-making. In case the consumer preferred a different SBoM format (say SPDX), they may have also chosen the tool to support the Translate property, to perform SBoM format conversion. Based on the different roles an organization performs, they should choose a tool that best suits their needs. For instance, the vendor could choose CycloneDX Generator (AppThreat, n.d.) to generate CycloneDX SBoMs and SCANOSS (SCANOSS, n.d.) or ORT (ORT, n.d.) if additional Analyze or Consume properties are desired. Similarly, the consumer of the software could have chosen OWASP's Dependency-Track (OWASP DT, n.d.), which provides component and known vulnerability analysis, license evaluation, and component identification platform.

Future Directions

Executive Order 14028 (NIST 2021) requests enhanced cybersecurity through maintaining privacy, security, and integrity of the software supply chain. NTIA is still in the process of formalizing records, and documentation relevant to SBoMs. In 2022 the National Institute of Standards and Technology (NIST) will be releasing recommendations to enhance software supply chain security along with issuing policies and standards (NIST 2021).

Barriers

Unfortunately, current SBoM usage in this sector is low due to the limited understanding of available tools and lack of standardization. Another barrier to establishing broader SBoM usage is the lack of a common platform to distribute and exchange SBoMs. Moreover, these SBoMs may contain sensitive or proprietary information, strongly necessitating the need of a privacy-preserving SBoM distribution and access platform. Also, for OT devices, there might exist a number of unknown subcomponents of software since the codebase may be old and the owner themself might be unaware of certain libraries or dependencies in the software product. Further, it is important for SBoM tools to handle dynamic and indirect dependencies. Since software nowadays is complex and cross-organizational, an SBoM tool should therefore be scalable before it is put into practice. Clearly, this domain is in a nascent stage requiring attention both from a research and development perspective.

Below the authors aim to pave a path forward, discussing some of the current issues that hinder the wide usage and adoption of SBoMs, highlighting the present needs, and providing recommendations to different actors associated with SBoMs.

Recommendations to Tool Developers, Researchers, and Authors

We identify here the following open challenges concerning SBoM tools.

- **Privacy and Security**. Current implementations of tools that enable sharing of SBoMs introduce problems encompassing sharing crucial and sensitive (sometimes confidential) information in a secure and privacy-preserving approach (Unisys, n.d.). One obstacle that a vendor may encounter is with sharing security-relevant information about a proprietary subcomponent produced by a third-party and contained in the vendor's software. Sharing software vulnerabilities in the digital supply chain may provide a malicious entity access to information before the vulnerability is patched. One of the challenges in this domain would be to check for the pres. ence of software components (and subcomponents) or vulnerabilities anonymously (i.e., without revealing the verifier's identity to the BoM owner). A privacy-preserving BoM query would prevent information leaks about components the verifier intends to use. Another possible challenge would be how a party could trust an SBoM or HBoM entry for the BoM's correctness and completeness (BoM attestation).

- ***Transform Tools***. CycloneDX CLI (CLI, n.d.) can convert from CycloneDX 1.2 to SPDX 2.1 and 2.2. For SPDX-based transformation, there exists some language-specific libraries (SPDX Golang, n.d.; SPDX BT, n.d.; SPDX Python n.d.), but the area still lacks a sophisticated tool that provides a wider range of format conversions without loss of information. Such tools are beneficial in instances where participating organizations adopt different SBoM formats.

- ***Recursive Verification***. Another possible area of research could be recursively verifying an SBoM. It is worth noting that modern software usually reuses third-party open-source code as subcomponents, thereby making SBoMs multi-level and complex. Attesting SBoMs, which may list a chain of software subcomponents, entails backtracing them in the complex supply chain. Normalizing this information still needs to be further studied and understood.

- ***Interoperability***. Different organizations, vendors, and asset owners sharing a supply chain may use different tools to generate or consume SBoMs. In such a situation, it is important to automatically identify a software subcomponent among multiple tools, systems, and databases. This can be achieved by using a consistent and unique primary key. A universal method to generate a complete identifier for a software component without the need for any extra information or central authority is necessary. CPE, one of the existing approaches in this domain, is a means of defining and distinguishing software including applications and operating systems, as well as hardware devices (Sanguino and Uetz 2017). The disadvantage of using CPE is that the naming convention is not unique, therefore composing a central dictionary is impractical (Fitzgerald and Foley 2013). Some other efforts in the area include the use of package URL and software heritage ID, which have still not been universally adopted (NTIA 2021f; PURL, n.d.). Another recommendation for organizations like Internet Engineering Task Force (IETF), and Internet Society (ISOC), is to standardize SBoM generation by identifying essential fields across different formats.

Recommendations to Standardizing Organizations

Closing the gap between defining standards and their respective implementations can be a challenging task. In this subsection, the authors discuss future directions for organizations dealing with standardizing SBoMs.

- **SBoM Formats.** Potential improvements in the existing SBoM formats would be the inclusion of the status of unknown subcomponents that may exist in a particular software product or if no such components exist. Organizations may also identify essential fields to standardize SBoM generation across different formats.

- **Dependent Links and Subcomponents.** One of the main applications of SBoM tools is software composition analysis (SCA) which aims to identify open source software in a product thereby assessing its security, license compliance, and code quality and providing vulnerability patches. The current SPDX format does not specify the information on any known patched vulnerabilities or updates.

- **HBoM Integration.** In addition, the authors propose that hardware-specific fields, based on the device type, can be added to the already existing SBoM formats, such as CycloneDX. These potential fields could specify barcode formats (Universal Product Code or European Article Number), media access control (MAC) address, a timestamp, stock-keeping unit (SKU) number, global trade item number (GS1 GTIN, n.d.), global location number (GS1 GLN, n.d.), global individual asset identifier (GS1 GIAI, n.d.), global model number (GS1 GMN, n.d.) and so on. Additionally, based on the type of device, specialized identification field(s) can be supplemented. For instance, a unique device identification (UDI) system can be added to the HBoM to uniquely identify medical devices (Food and Drug Administration 2013).

- **SBoM Distribution and Access.** The goal is to have a BoM be accessible and available to the right entities at the right time. SBoMs of open-source components should be easily available for a consumer or a developer reusing the component to view through a global access point. Although SPDX provides a list of accepted licenses (SPDX-Licenses, n.d.), a common domain for SBoMs is still not available. DBoM (Unisys, n.d.) aims to improve attestation and BoM sharing across different entities sharing a supply chain but still requires a way to map different databases supporting relational or non-relational data formats. An SBoM may be distributed to the consumer through a URL using a manufacturer usage description (MUD) extension (NTIA, 2021b; IETF 2019). Another way to achieve this task is to provide a manifest offering licensing specifications and package contents to the downstream consumer of the software. In addition, a publish or subscribe system would prove to be advantageous, allowing consumers to get notified about updates from the supplier through a shared channel. In other words, a consumer can access the SBoM by getting it directly from the supplier, say through an email. In case the SBoM exists on the device executing the software to which it belongs, it can be fetched using protocols like HTTPS. An SBoM can also be made available on a shared repository.

Recommendations to Users: Choosing a Tool

Based on various properties of an SBoM tool, such as repository and format supported, user interface and most importantly its functionalities (i.e., produce, consume, and transform) a choice can be made. For instance, FOSSology (Gobeille 2008) and ORT (ORT, n.d.) are two of the most diverse open source SBoM tools available. Below the authors provide additional insight which may help users to select a tool as per their requirements.

- *Insight on SBoM Format.* It is worth noting that SWID tags currently support only the XML format which is perceived to have poor performance characteristics concerning its parsing (Nicola and John 2003). JSON files, on the other hand, are easier to consume. Using the lightweight CoSWID may therefore be beneficial in certain scenarios like usage in constrained devices. The authors propose the use of open standard- CycloneDX which can provide vulnerability details through the fields cpe, purl and swid. In addition, this lightweight format was specifically designed to improve application security and facilitate supply chain component analysis and therefore incorporates characteristics of SWID and SPDX formats. The format also has an active community and support.

- *Insight on Tool Repositories.* Software used across the globe handles sensitive information and carries out critical tasks which raise the importance of having clarity into the software's composition. The concept of building and exchanging BoMs of these contemporary devices can promise transparency into its components, defending their networks from known vulnerabilities, therefore, this information must be available to those who require it. It is worth noting that the BoM data is inherently relational as shown in Figure 2. This is due to the diverse nature of the contemporary supply chain architecture as well as code reuse, which is an integral part of modern software. In general, a relational database would be most appropriate for a BoM application as this would preserve the logical connections (i.e., relations between the data) (Goggins, Germonprez, and Lumbard 2021; Gobeille 2008; OWASP DT, n.d.; Tern, n.d.). A blockchain-based architecture that uses a relational database would also serve the purpose in case properties, such as decentralization and immutability, provided by blockchain technology are desired (OWASP DT, n.d.). A choice between various relational database tools can be made based on other desired properties like security, performance features, and so on (SParts, n.d.). The authors believe that PostgreSQL (Obe and Hsu 2017) would be an appropriate choice for an SBoM tool and would be beneficial in situations when an existing repository or vendor-native database is non-relational. As per the authors' knowledge, the only shortcoming of PostgreSQL is that it is a centralized database, which implies that the server storing the DBoM database is a single point of failure. To tackle such a situation, PostgreSQL allows for replica servers, maintaining database availability in case of a failover situation (PostgreSQL, n.d.). Blockchain technology has earned popularity over the past few years, which cannot be denied. But for the purpose of choosing a BoM repository, the authors argue that a blockchain-based database does not hold many benefits. The authors, therefore, propose to use a much simpler database like PostgreSQL.

Conclusion

In this paper, the authors systematically describe BoM terminologies, classifications, and the desired properties in SBoM tools. We further specify broad use cases of employing BoMs and discuss case studies, highlighting its benefits. In addition, the paper investigates advancement in the area of SBoMs, thereby helping readers to identify tools that best suit their requirements. Finally, we authors identify missing pieces in existing BoM implementations and provide recommendations to tool developers, users, researchers, and standardizing organizations, thereby, providing some pathways for future research.

Acronyms and Abbreviations

AI	Artificial Intelligence
API	Application Programming Interface
BoM	Bill of Materials
CPE	Common Platform Enumeration
DBoM	Digital Bill of Materials
DoS	Denial of Service
EOL	End-of-life
HBoM	Hardware Bills of Materials
IETF	Internet Engineering Task Force
ISOC	Internet Society
IT	Information Technology
JSON	JavaScript Object Notation
MAC	Media Access Control
MUD	Manufacturer Usage Description
NIST	National Institute of Standards and Technology
NTIA	National Telecommunication and Information Administration
OT	Operational Technology

OWASP	Open Web Application Security Project
SBoM	Software Bill of Materials
SCA	Software Composition Analysis
SKU	Stock-Keeping Unit
SPDX	Software Package Data Exchange
SWID	Software Identification
UDI	Unique Device Identification
XML	Extensible Markup Language

Author Capsule Bios

Arushi Arora is a Ph.D. student (majoring in Computer Science) and a Research Assistant under Dr. Christina Garman at Purdue University. She also jointly works with National and Homeland Security at Idaho National Laboratory. Her research interests include information security, anonymity networks, and applied cryptography. She received her bachelor's in Computer Science in 2019 from Indira Gandhi Delhi Technical University, India, where she was awarded the Chancellor's and Vice Chancellor's Gold Medal for her academic excellence.

Virginia "Ginger" Wright is the Energy Cybersecurity Portfolio Manager for Idaho National Laboratory's Cybercore division within its National and Homeland Security directorate. She leads programs focused on cybersecurity and resilience of critical infrastructure for DOE, DARPA and other government agencies including DOE's CyTRICS™ program. Ms. Wright's recent research areas include supply chains for operational technology components, incident response, critical infrastructure modeling and simulation, and nuclear cybersecurity. Ms. Wright has a Bachelor of Science in Information Systems/Operations Management from the University of North Carolina at Greensboro.

Christina Garman is an assistant professor in the Department of Computer Science at Purdue University. Her research interests focus largely on practical and applied cryptography, namely the design and analysis of real-world cryptographic systems. She aims to make it easier to design and securely deploy new and complex cryptographic systems while preventing insecurities from occurring in such systems. She received an NSF CAREER Award in 2021, and her work has received a best paper award at ACM CCS and been featured in numerous media, including The Washington Post, The New York Times, Wired, and The Economist. She received her MS and Ph.D. from Johns Hopkins University in Computer Science in 2013 and 2017 respectively, and a BS in Computer Science Engineering and a BA Mathematics, with a minor in Physics, from Bucknell University in 2011.

References

Aboutorabi, Seyyed Hamid, Mehdi Rezapour, Milad Moradi, and Nasser Ghadiri. 2015. "Performance evaluation of SQL and MongoDB databases for big e-commerce data." *In 2015 International Symposium on Computer Science and Software Engineering (CSSE)*, 1–7. IEEE. https://doi.org/10.1109/CSICSSE.2015.7369245.

AppThreat. n.d. "CycloneDX Generator." Accessed Jan 18, 2022. https://github.com/AppThreat/cdxgen.

Bauer, Andreas, Nikolay Harutyunyan, Dirk Riehle, and Georg-Daniel Schwarz. 2020. "Challenges of tracking and documenting open source dependencies in products: A case study." *Open Source Systems* 582, 25-35. https://dx.doi.org/10.1007%2F978-3-030-47240-5_3.

Boicea, Alexandru, Florin Radulescu, and Laura Ioana Agapin. 2012. "MongoDB vs Oracle– database comparison." *In 2012 third international conference on emerging intelligent data and web technologies*, 330–335. IEEE. https://doi.org/10.1109/EIDWT.2012.32.

CISA. 2021. "Mitigating Log4Shell and Other Log4j-Related Vulnerabilities". Accessed March 31, 2022. https://www.cisa.gov/uscert/ncas/alerts/aa21-356a.

CLI. n.d. "CycloneDX CLI." Accessed Jan 18, 2022. https://github.com/CycloneDX/cyclonedx-cli.

CSIS. 2022. "Executive Order 14028 and Federal Acquisition Regulation (FAR): 10 Months Later". Accessed June 29, 2022. https://www.csis.org/blogs/strategic-technologies-blog/executive-order-14028-and-federal-acquisition-regulation-far-10.

CycloneDX. n.d. "CycloneDX Specifications." Accessed Jan 18, 2022. https://cyclonedx.org/specification/overview/.

Dwyer, AC. 2018. "The NHS cyber-attack: A look at the complex environmental conditions of WannaCry." *RAD Magazine* 44.

Fitzgerald, William M, and Simon N Foley. 2013. "Avoiding inconsistencies in the security content automation protocol." *In 2013 IEEE Conference on Communications and Network Security (CNS)*, 454–461. IEEE. https://doi.org/10.1109/CNS.2013.6682760.

Food and Drug Administration, HHS. 2013. "Unique device identification system. Final rule." *Federal register* 78, no. 185 (2013): 58785-58828. https://www.federal

register.gov/d/2013-23059.

Gandhi, Robin, Matt Germonprez, and Georg JP Link. 2018. "Open data standards for open source software risk management routines: an examination of SPDX." *In Proceedings of the 2018 ACM Conference on Supporting Groupwork*, 219–229. https://doi.org/10.1145/3148330.3148333.

Gobeille, Robert. 2008. "The fossology project." In Proceedings of the 2008 international working conference on Mining software repositories, 47–50. https://doi.org/10.1145/1370750.1370763.

Goggins, Sean P, Matt Germonprez, and Kevin Lumbard. 2021. "Making Open Source Project Health Transparent." *Computer* 54 (08): 104–111. https://doi.ieeecomputersociety.org/10.1109/MC.2021.3084015.

Greenberg, Andy. 2018. "The untold story of NotPetya, the most devastating cyberattack in history." *Wired, August 22*. https://www.wired.com/story/notpetya-cyberattack-ukraine-russia-code-crashed-the-world/.

GS1 GIAI. n.d. "Global Individual Asset Identifier (GIAI)." Accessed Jan 19, 2022. https://www.gs1.org/standards/id-keys/global-individual-asset-identifier-giai.

GS1 GLN. n.d. "Global Location Number (GLN)." Accessed Jan 19, 2022. https://www.gs1us.org/DesktopModules/Bring2mind/DMX/Download.aspx?Command=Core_Download&EntryId=158&language=en-US&PortalId=0&TabId=134.

GS1 GMN. n.d. "Global Model Number (GMN)." Accessed Jan 19, 2022. https://www.gs1au.org/what-we-do/standards/global-model-number-gmn.

GS1 GTIN. n.d. "Global Trade Item Number (GTIN)." Accessed Jan 19, 2022. https://www.gs1.org/standards/id-keys/gtin.

IETF. 2019. "Manufacturer Usage Description Specification." https://datatracker.ietf.org/doc/html/rfc8520.

Jatana, Nishtha, Sahil Puri, Mehak Ahuja, Ishita Kathuria, and Dishant Gosain. 2012. "A survey and comparison of relational and non-relational database." *International Journal of Engineering Research & Technology* 1 (6): 1–5.

Li, Yishan, and Sathiamoorthy Manoharan. 2013. "A performance comparison of SQL and NoSQL databases." *In 2013 IEEE Pacific Rim Conference on Communications, Computers and Signal Processing (PACRIM)*, 15–19. IEEE. https://doi.org/10.1109/PACRIM.2013.6625441.

Liu, Min, Jianbo Lai, and Weiming Shen. 2014. "A method for transformation of engineering bill of materials to maintenance bill of materials." *Robotics and Computer-Integrated Manufacturing* 30 (2): 142–149. https://doi.org/10.1016/j.rcim.2013.09.008.

Nicola, Matthias, and Jasmi John. 2003. "Xml parsing: a threat to database performance." *In Proceedings of the twelfth international conference on Information and knowledge management*, 175–178. http://dx.doi.org/10.1145/956863.956898.

NIST. 2021. "Improving the Nation's Cybersecurity: NIST's Responsibilities under the May 2021 Executive Order." Accessed Jan 18, 2022. https://www.nist.gov/itl/executive-order-improving-nations-cybersecurity.

NTIA. 2021a. "Framing Software Component Transparency: Establishing a Common Software Bill of Material (SBOM)." https://www.ntia.gov/files/ntia/publications/framingsbom_20191112.pdf.

NTIA. 2021b. "Sharing and Exchanging SBOMs." https://www.ntia.gov/files/ntia/publications/ntia_sbom_sharing_exchanging_sboms-10feb2021.pdf.

NTIA. 2021c. "Taxonomy, SBoM Tool Classification." https://www.ntia.gov/files/ntia/publications/ntia_sbom_tooling_taxonomy-2021mar30.pdf.

NTIA. 2021d. "Software Bill of Materials." https://www.ntia.gov/SBOM.

NTIA. 2021e. "Roles and Benefits for SBOM Across the Supply Chain." https://www.ntia.gov/files/ntia/publications/ntia_sbom_use_cases_roles_benefits-nov2019.pdf.

NTIA. 2021f. "Software Identification, and Guidance Challenges." https://www.ntia.gov/files/ntia/ publications/ntia_sbom_software_identity-2021mar30.pdf.

NTIA. 2021g. "The Minimum Elements For a Software Bill of Materials (SBoM)." https://www.ntia.doc.gov/files/ntia/publications/sbom_minimum_elements_report.pdf.

Obe, Regina O, and Leo S Hsu. 2017. "PostgreSQL: Up and Running: A Practical Guide to the Advanced Open Source Database". *O'Reilly Media, Inc.*

ORT. n.d. "OSS Rewiew Toolkit". Accessed Jan 18, 2022. https://github.com/oss-review-toolkit.

OWASP DT. n.d. "OWASP's Dependency-Track". Accessed Jan 18, 2022. https://

docs.dependencytrack.org.

Parnas, David Lorge, GJK Asmis, and Jan Madey. 1991. "Assessment of safety-critical software in nuclear power plants." *Nuclear safety 32 (2): 189–198.*

PostgreSQL. n.d. "PostgreSQL Replication & Automatic Failover Tutorial". Accessed Jan 18, 2022. https://www.enterprisedb.com/postgres-tutorials/postgresql-replication-and-automatic-failover-tutorial#replication-use.

PURL. n.d. "Package URL Specifications." Accessed Jan 18, 2022. https://github.com/package-url/purl-spec.

Sanguino, Luis Alberto Benthin, and Rafael Uetz. 2017. "Software vulnerability analysis using CPE and CVE." *arXiv preprint arXiv*:1705.05347.

SCANOSS. n.d. "SCANOSS." Accessed Jan 18, 2022. https://scanoss.com.

Shiroudi, A, R Rashidi, GB Gharehpetian, SA Mousavifar, and A Akbari Foroud. 2012. "Case study: Simulation and optimization of photovoltaic-wind-battery hybrid energy system in Taleghan-Iran using homer software." *Journal of Renewable and Sustainable Energy* 4 (5): 053111. http://dx.doi.org/10.1063/1.4754440.

Singhal, Anjali, and RP Saxena. 2012. "Software models for smart grid." *In 2012 First Interna- tional Workshop on Software Engineering Challenges for the Smart Grid (SE-SmartGrids),* 42–45. IEEE. https://doi.org/10.1109/SE4SG.2012.6225717.

Slootweg, JG, SWH De Haan, H Polinder, and WL Kling. 2003. "General model for representing variable speed wind turbines in power system dynamics simulations." *IEEE Transactions on power systems* 18 (1): 144–151. https://doi.org/10.1109/TPWRS.2002.807113.

SParts. n.d. "Sparts Documentation." Accessed Jan 18, 2022. https://sparts.readthedocs.io/en/latest/web/intro.html.

SPDX BT. n.d. "SPDX Build Tool." Accessed Jan 18, 2022. https://github.com/spdx/spdx-build-tool.

SPDX Golang. n.d. "SPDX Golang Library." Accessed Jan 18, 2022. https://github.com/spdx/tools-golang.

SPDX Python. n.d. "SPDX Python Library." n.d. Accessed Jan 18, 2022. https://github.com/spdx/tools-python.

SWID Tags. n.d. "Guidelines for the Creation of Interoperable Software Identification (SWID) Tags." Accessed Jan 18, 2022. https://nvlpubs.nist.gov/nistpub.

Unisys. n.d. "Digital Bill of Materials." Accessed Jan 18, 2022. https://dbom-project.readthedocs.io/en/latest/what-dbom.html.

Waltermire, David, and Brant Cheikes. 2015. "Forming common platform enumeration (CPE) names from software identification (SWID) tags." *Technical report. National Institute of Standards and Technology.* https://csrc.nist.gov/publications/detail/nistir/8085/draft.

Wolff, Evan D., K. M. Growley, and M. G. Gruden. 2021. "Navigating the solarwinds supply chain attack." *The Procurement Lawyer* 56, no. 2. https://www.crowell.com/files/20210325-Navigating-the-SolarWinds-Supply-Chain-Attack%20.pdf.

Zhou, Chunliu, Xiaobing Liu, Fanghong Xue, Hongguang Bo, and Kai Li. 2018. "Research on static service BOM transformation for complex products." *Advanced Engineering Informatics* 36:146–162. https://doi.org/10.1016/j.aei.2018.02.008.

A Risk-Informed Community Framework for the Assessment of Chemical Hazards

Curtis Smith,[1] Kurt Vedros,[2,3] Kenneth Martinez,[4] David Kuipers[5]

[1] Director, Nuclear Safety and Regulatory Research Division, Idaho National Laboratory

[2] Lead Risk Assessment Engineer, Nuclear Science and Technology Division, Reliability, Risk and Resilience Sciences Department, Idaho National Laboratory

[3] Corresponding Author, kurt.vedros@inl.gov

[4] Senior Critical Infrastructure Analyst, Idaho National Laboratory

[5] Critical Infrastructure Cybersecurity Engineer, Idaho National Laboratory

Abstract

We present the Data and Risk-Informed Chemical Assessment Technique (DRICAT), a quantitative/qualitative risk analysis technique for assessing the risk of a potential chemical release incident that may lead to a mass casualty event in United States communities. Risk assessment is a comprehensive, structured, and logical analysis approach aimed at identifying and assessing risks in "systems" for the purpose of improving management of these systems. DRICAT leads to better understanding and effective management of risks from chemical incidents through risk and scenario identification and ranking by severity by helping community planning to minimize morbidity and mortality during and after potential events. As such, DRICAT is designed to be reproducible, evidence-based, practical, and scalable for different types of communities and the possible chemical hazards present in that community. Recognizing that many communities have assessment protocols and response mechanisms already in place, we believe these DRICAT characteristics will enhance both existing chemical incident awareness and readiness activities while providing a baseline approach for communities lacking chemical release risk analysis techniques.

To understand where potential hazards might exist within a community, it is useful to consider drivers for hazards (i.e., those factors that influence the presence of the hazards and the uncertainty). DRICAT uses essential elements of information (EEI) to identify chemical initiating events (IE) which are a part of potential hazardous scenarios. Both formal and informal approaches can be used to identify initiators arising from chemical hazards. DRICAT focuses on precursor events and a deductive approach using a hazard identification diagram. EEI data sources for community chemical hazards range from informal (e.g., social media, and local news outlets) to formal vetted databases. EEIs include systematic identification of hazards, community factors that affect hazards (e.g., population density, weather, and commodity flows), associated IEs, and grouping of individual causes into like categories. IE characteristics may vary among communities and include IEs that may lead directly to a chemical release or may require additional mitigative failures. DRICAT leverages EEIs to identify IEs to build and rank chemical accident scenarios from IE to the potential outcome. The likelihood of this release and the consequence of the release determine the overall risk.

Keywords: Chemical risk assessment, Community chemical hazard profile, Quantitative risk assessment, Qualitative risk assessment, Hazardous material accident assessment, Chemical terrorism

Introduction

Risks exist in many ways. While ranging in degree from personal hazards to the potential severe consequences associated with complex technological systems, risk nonetheless is embodied in a variety of situations and conditions. When asked, the layperson can generally provide information related to his or her perception of risk. But what exactly is risk? And, more importantly, what operation definition do we attach to risk when performing or using a risk assessment? The U.S. Department of Homeland Security (DHS) definition of risk is "potential for an unwanted outcome resulting from an incident, event, or occurrence, as determined by its likelihood and the associated consequences" [DHS 2010]. This definition identifies the essential parameters which are necessary to define risk. In brief, risk is the chance of experiencing a specified set of undesired consequences.

Risk assessment is a comprehensive, structured, and logical method aimed at identifying and assessing risks in "systems" for the purpose of improving mitigation effectiveness. As part of the technique we propose, we wish to better understand and effectively manage high impact community chemical risks to protect the public, identify these risks, and minimize morbidity and mortality during and after potential events. As such, our "system" for risk assessment purposes is any community under consideration and the possible high impact hazards present in that community.

Hazardous chemicals exist in most communities today in varying quantities and toxicity. All communities have business entities who use potentially hazardous chemicals in their day-to-day operations. This includes businesses that manufacture chemicals or may use hazardous chemicals in manufacturing processes. These chemicals need to be transported to or from a site that uses or manufactures them. As a result, communities have transportation routes that serve as conduits for moving the hazardous chemicals into, out of, and through their boundaries. The potential chemical hazards and the risks that they pose to the occupational workforce and surrounding regional population should be better understood for each region within a jurisdiction allowing the community to plan and improve preparedness for potential emergency events. A risk assessment process can support a community in enhancing awareness of their hazardous chemical preparedness level and can be used to inform emergency agencies and first responders.

Definition of Risk and Risk Assessment

The general concept of risk includes undesirable scenarios and their corresponding consequences, and likelihoods, (e.g., the number of people harmed, and the probability of occurrence of this harm). Historically, risk is represented as a set of questions [Kaplan, Stanley, and B. John Garrick 1981. 11-27]:

- What can go wrong? This question implies the need to understand a possible hazard and associated potential accident scenarios.

- How likely is it to occur? This question implies the need to understand how possible it is to experience a possible hazard and associated scenarios.

- What will be the outcome? This question implies the need to understand how impacts to the public and critical infrastructure (CI) may be realized (i.e., consequences).

In addition to this classic example, we add:

- What are the uncertainties of qualitative and/or quantitative approaches to determining risk? This question implies the need to understand how much confidence we have in our understanding of possible hazards, associated scenarios, and potential consequences.

Hazards, scenarios, and uncertainties are among the most important components of a risk assessment. Figure 1 shows the implementation of these concepts in a risk assessment. In this Figure 1, uncertainty is shown to be an integral part of each step of the process.

The accident scenarios begin with a hazard that may present a set of "initiating events" (IEs) that perturb a community. These IEs are upset conditions that start a potential chemical-related scenario. Note that impacts (consequences) could include items such as population health impacts, business interruption impacts, environmental impacts, etc. However, we are only considering a health impact to the population representing a community. A common element in our understanding of chemical risks is the idea of "hazards." A hazard is a condition that is, or potentiates, a deviation that is undesired.

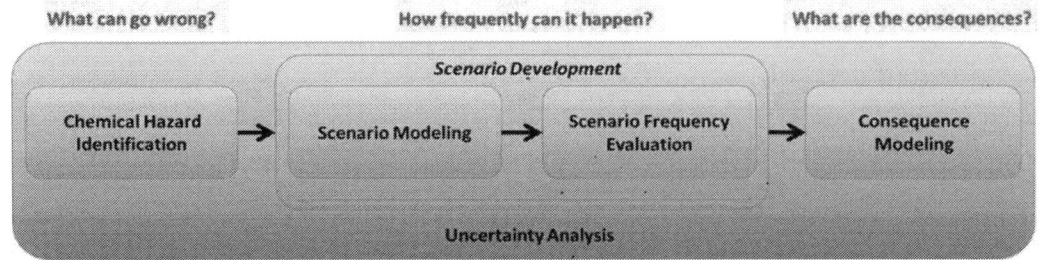

Figure 1. Implementation of the scenario concept in risk assessment.

To understand where potential hazards might exist within a community, it is useful to consider "drivers" for hazards. Hazard drivers are the factors that influence the presence of a hazard and its uncertainty. For example, the interstate highway and rail lines that pass through a city brings needed connectivity and commerce, but also bring a chemical transportation accident hazard. More subtle drivers can be culture, local ordinances, and work practices.

Not all deviations from a desired condition in the community will lead to mishaps or accidents; however, if operational intent is understood appropriately, all mishaps and accidents will be found to correspond to deviations from desired operational goals. Therefore, specification of operational intent implies different categories of potential chemical hazards and provides cues to an identification process. Further, these hazards typically are considered for the various safety impacts shown in Figure 2, where two fundamental types of impacts are present when considering chemical hazards, specifically physical or biological impacts.

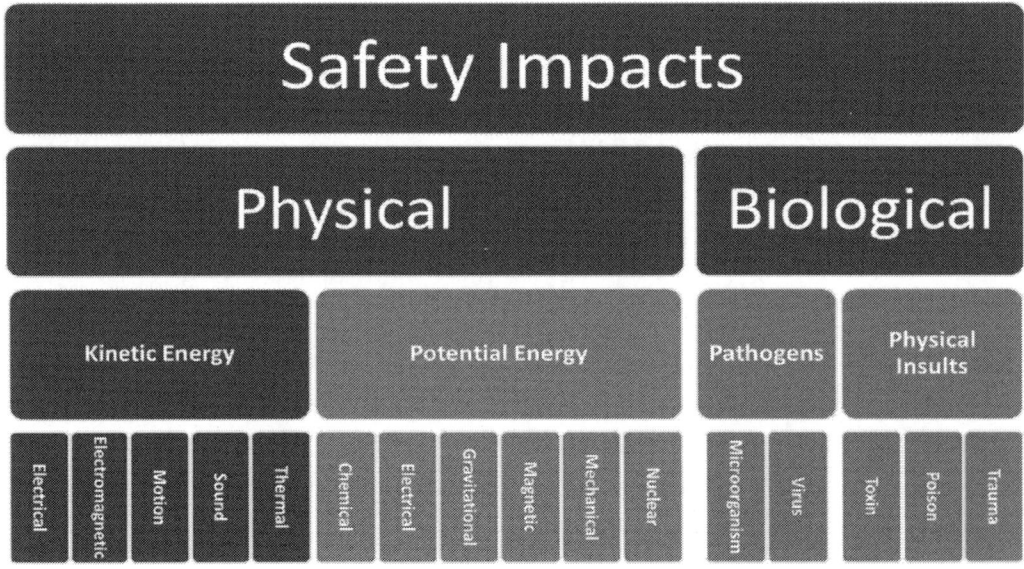

Figure 2. Hazard characteristics related to chemical risk assessment.

In general, health hazards can be represented by one or more of the characteristics found in Figure 2, including:

- Kinetic energy. These include chemical based explosions that result in excessive motion (e.g., missiles and overpressure), sound (e.g., overpressure), and thermal (e.g., fire) impacts.

- Potential energy. These include chemicals that result in stored energy such as hydrogen, liquefied natural gas, gasoline, etc.

- Pathogens. These contain organisms (e.g., virus, bacterium, protozoan, and fungus) that may cause a disease.

- Physical insults. These include chemicals that are a toxin, poison, or cause direct trauma. The hazard may result in short-, intermediate-, or long-term impacts. For example, an ammonia leak could kill immediately, while a ground water leak of jet fuel storage can be linked to long-term cancer deaths.

All of the hazard characteristics listed may be described in terms of their impact or volatility and toxicity. By using a holistic risk-informed approach, we can focus on community-level hazards that could include both chemical and biological impacts. However, the focus of this paper will be on those hazards that are precipitated by chemicals that may be found in communities across the U.S.

Plausible scenarios are investigated once an exhaustive list of hazards and their characteristics are identified. The starting point for a scenario is the initial perturbation of the hazard (the IE). For each IE, we typically represent a deviation that leads to a scenario's evolution. A chemical-related scenario may lead (or not) to an undesired outcome such as a chemical release into a population of people. As part of the scenario representation, mitigating features are considered, if present, to capture realistic aspects of the scenario. Also, scenario modeling may be either qualitative, quantitative, or a mixture of both. Lastly, multiple hazards and multiple potential scenarios are integrated together. These, with an understanding of the uncertainties present, create a risk picture for chemical hazards within a community. This risk picture then supports risk management such as prioritization, mitigative measures, and minimization of consequences via proactive action.

A risk scenario is defined by DHS as a "hypothetical situation comprised of a hazard, an entity impacted by that hazard, and associated conditions including consequences when appropriate" [DHS 2010]. The concept of a community chemical risk scenario is illustrated in Figure 3. A scenario represents the context from: (1) a chemical hazard that may be present in a community, (2) the starting point (IE) that initiates the scenario, and (3) the failure of mitigative hazard prevention approaches. Combined, these three elements provide a way to determine the frequency of a potential accident and the severity of its associated consequences.

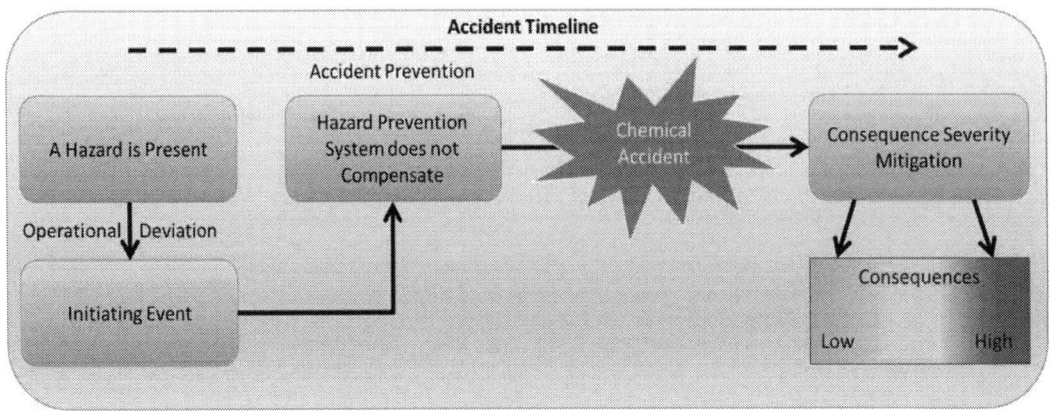

Figure 3. The concept of a scenario.

Communities are at risk to chemical hazards and prepare for the hazards they determine to be likely and impactful. While nearly all communities have processes for the assessment and response to chemical accidents and some are quite sophisticated, far too often, decisions regarding which hazards to prepare for are often made by solely interviewing subject matter experts and not driven by data or a combination of both. This can lead to mis-identifying the most impactful hazards and/or scenarios. The discussion that follows presents the Data and Risk-Informed Chemical Assessment Technique (DRICAT), a risk assessment methodology to determine the chemical accident risk present in a community that can improve existing assessment processes and reduce the chance of mis-appropriating critical resources. In addition, DRICAT is not intended to replace organizations that respond to hazardous events but is intended to support the identification of appropriate federal datasets (e.g., EPA or DHS) and to create a systematic approach for collating the data in a useable format for emergency managers.

Risk assessment is a process that utilizes hazards assessment and accident scenario analyses introduced above to provide a complete assessment of risk. In summary, risk assessment is a process that has the following characteristics:

- Follows a rigorous, systematic approach that requires information integration (multidisciplinary)

- Uses a qualitative and/or quantitative picture of risks associated with performance measures

- Captures dependencies and other (inter/intra)-relationships

- Includes human and system elements

- Helps to balance efforts between prevention and mitigation

- Works within a scenario-based approach that informs decision-making

- Identifies contributing elements

- Helps to prioritize (focus on where improvements will be effective)

- Provides a framework for monitoring and trending

Methods: Data and Risk-Informed Chemical Assessment Technique (DRICAT)

To assess a chemical hazard in a community and provide a technically defensible approach to risk management, we have proposed DRICAT. This is a process that has the following attributes:

- Follows a simple process for state/local decision-makers

- Enhances existing assessment processes

- Is based on a risk assessment process that is reproducible

- Is informed by local data such as potential community chemical hazards, precursor events, and local information from hazmat sources such as law enforcement and fire departments

- Will allow for scalability between cities of different size and different scenarios of interest

- Has a goal to inform emergency planners of priority chemical hazards to focus planning measures to reduce community impacts

For DRICAT, we will primarily focus on the planning stages which are shown in the left portion of Figure 3, specifically the identification of chemical hazards in a community, how they might impact a community through scenarios, and the likelihood of an event leading to non-negligible consequences in a community. These steps are shown as Steps 1–3 in Figure 4.

We will describe each of the three steps in detail. Note that for a community-level risk assessment using DRICAT, we will require novel approaches to data collection, interpretation, and application.

Critical to the DRICAT process are the Essential Elements of Information (EEI) defined by the Federal Emergency Management Agency (FEMA) [FEMA 2013] as "essential information requirements that are needed for informed decision making." The key EEIs identified in Figure 5, through research and identification can be collected, collated, and will create the data streams to be used to build the community chemical risk profile by using the DRICAT process. We will address as a component of known chemical inventory risks, accidents and acts

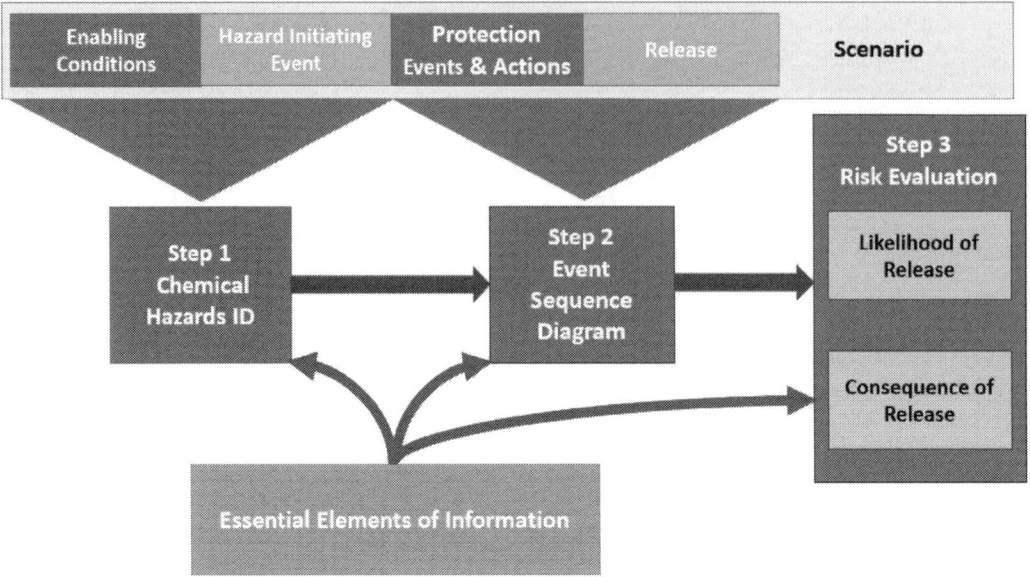

Figure 4. The three key steps of DRICAT.

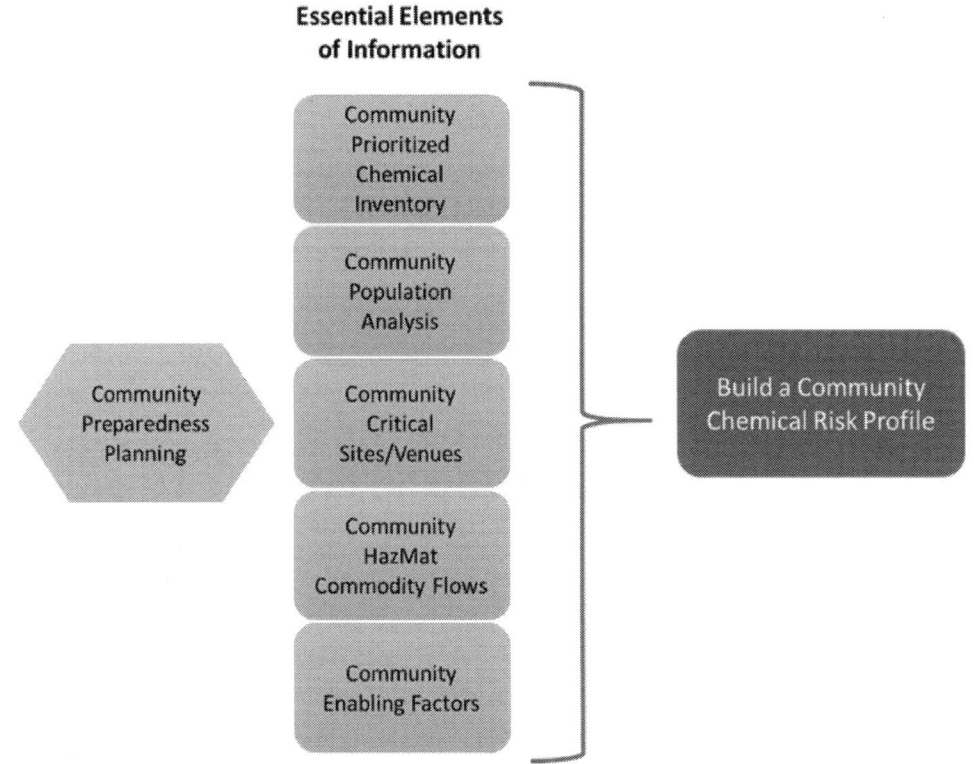

Figure 5. EEIs used in DRICAT.

of terrorism such as sabotage or cyber-attack to regional chemical production or storage vessels. Focusing on the known chemical risk hazards first creates an atmosphere of awareness, understanding, and communication across all those with roles and responsibilities for responding to a chemical incident. Planning, training, and exercising for the known chemical risk hazards ensures that all those involved in a chemical incident response will operate efficiently and collaboratively on incidents both planned and unplanned.

In addition to the many specific sources of information listed throughout this article, the authors note that it is useful to review more general information from programs of the DHS [DHS 2015], Cybersecurity & Infrastructure Security Agency [CISA, 2022], and the Environmental Protection Agency [EPA 1999], [EPA 2022]. It should also be noted that while much information has been collected related to chemical hazards in the U.S., several issues remain including integration of this information into a coherent picture of relative risks, aging of the data sources, and the vastness of the data sources cause difficulties in finding high-value information.

Chemical Hazards Identification

Through active, a priori identification of the chemical inventory in a region, communities become strategically aware of the known chemical risk hazards that could impact a regional population should the chemical be released. Identification should include an understanding of specific factors, including toxicity, quantity, and volatility for each hazardous chemical. These factors contribute to a risk prioritization that focuses planning and preparedness efforts to the highest risk hazardous chemical agents. Higher toxicity agents present more significant health impacts to a regional population that could occur in shorter periods of exposure. The greater the quantity of a hazardous chemical agent that is produced and/or stored results in increases in the numbers of a regional population that could be impacted should the chemical be released. Volatility focuses on those chemical agents that are airborne. Inhalation of a toxic chemical presents the most significant risk of morbidity and mortality since the chemical is introduced directly into the blood stream once it reaches the alveoli of the lungs.

Various approaches exist for identification of IEs, including the use of master logic diagrams (MLDs). A MLD is a hierarchical, top-down display of IEs, showing general types of undesired events at the top, proceeding to increasingly detailed event descriptions at lower tiers, and displaying IEs at the bottom. Figure 6 shows a representative MLD for a chemical leak with a production tank fueling leak accident drilled down to the IE level. The goal is not only to support identification of a comprehensive set of IEs but also to group them according to the challenges that they pose (the responses that are required because of the attributes of their occurrences).

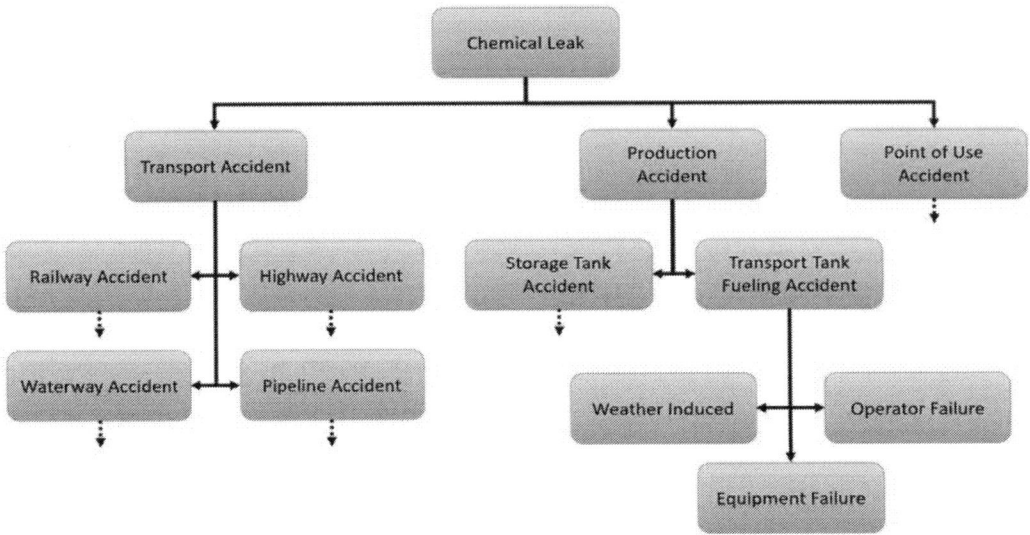

Figure 6. Example of a MLD for a chemical leak.

The depiction of initiators comes from different techniques. Both formal and informal approaches can be used to identify initiators arising from hazards, such as:

- Precursor events may indicate the types and frequencies of applicable upsets

- Ad hoc methods such as surveys and expert elicitation may provide local hazards that are known or have occurred in the past

- Deductive approaches such as failure modes and effects analysis, fault trees, hazard identification diagrams (HID), and MLDs may be used.

An example of a HID is shown in Figure 7 where key characteristics of the chemical hazard are questioned to keep or screen out a specific hazard. These characteristics include:

- The properties of the chemical hazard – high level of toxicity (a biological insult), volatility, or detonation potential (a kinetic energy insult) passes the hazard to the next level of screening

- The quantity of the chemical – a high enough quantity to maintain the potential high-risk passes the hazard to the next level of screening while lower quantities are held for possible inclusion

- The containment environment used to store and protect the chemical – low quality or compromised containment will pass the hazard to consideration for inclusion in scenarios, high quality containment can screen the hazard, and medium quality containment causes the hazard to be held for possible inclusion

- The questioning process continues, using all the applicable characteristics (EEIs) collected (e.g., the local terrain near the chemical and local population density near the chemical).

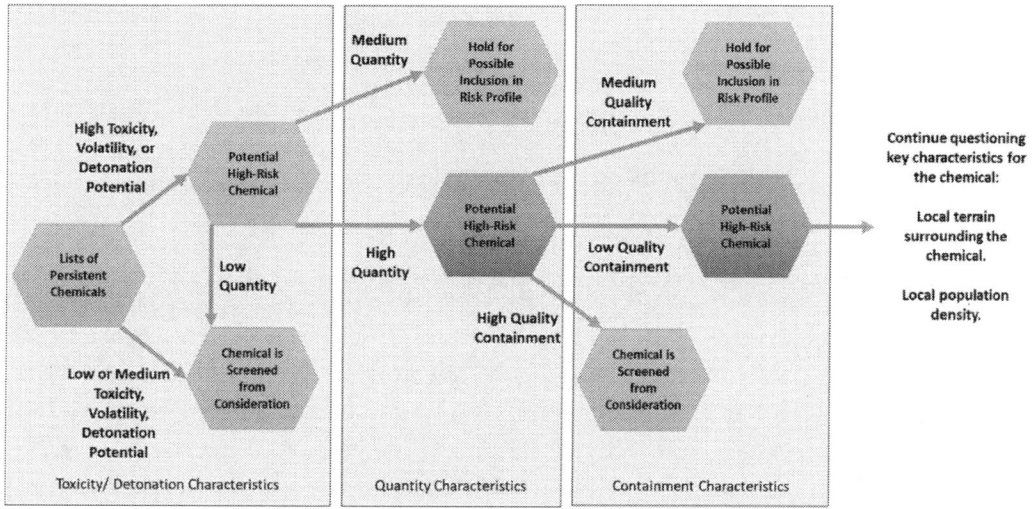

Figure 7. Example HID for identifying potential chemical hazards.

Data sources for community chemical hazards range from informal (e.g., social media, local news outlets) to formal vetted databases such as the EnviroAtlas hosted by the Environmental Protection Agency [EPA 2022]. Figure 8 shows an example of the EnviroAtlas graphical database listing local Superfund, toxic release inventory, and hazardous waste sites.

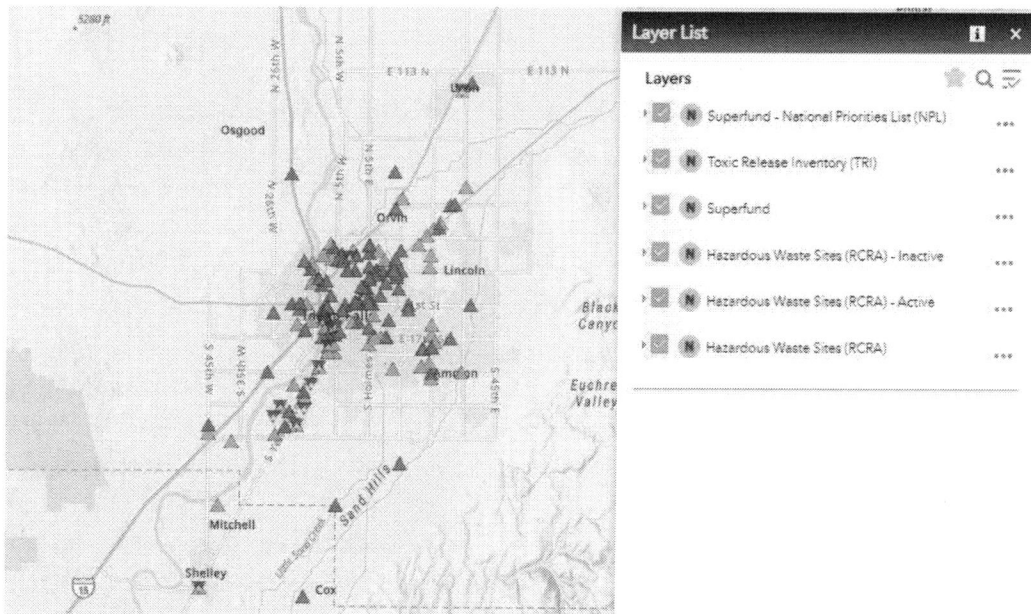

Figure 8. EnviroAtlas database showing potential chemical hazards (EPA 2022).

Community Chemical Inventory

This area focuses on the chemical data EEI needed to evaluate hazard potential for each entity in or near a community which could impact safety of community population due to a chemical release. Chemical inventory sources and information include:

- Rules and regulations related to hazardous chemicals in the community

- Existing data sets that provide community level data

- Chemical properties – Toxicity and Volatility.

The approach is to consult federal, state, and local rules and regulations to determine how hazardous chemicals (HC) should be produced, transported, and handled within the community. This step looks at collecting an easily referenced list of HC data sets for planning and emergency use. These data sets are available through producers, users, and federal, state, county, and city information (the list is too extensive to address in this paper). Data is provided on locations, amounts and points of use for HC. The jurisdiction of these data sets and their public availability varies. Issues such as business sensitive proprietary information need to be balanced with the need for public safety planning and response.

Under the Emergency Planning and Community Right-to-Know Act (EPCRA), Congress requires each state to appoint a State Emergency Response Commission (SERC), which divides their state into Emergency Planning Districts. Each district must develop a Local Emergency Planning Committee (LEPC) which is made up of representatives from state and local officials, emergency response and public health, environment, transportation, hospital, industry, and community group and media entities. The LEPC must develop an emergency response plan, review the plan at least annually, and provide information about chemicals in the community to citizens. EPCRA regulations require entities submit hazardous chemical information to their district LEPC and local fire department and hazmat units. The local fire department and hazmat response units should be consulted for additional relevant information about local HCs obtained through inspections and can share this information as key members of the planning and response team.

Ranking chemicals by the severity of the hazard they present requires an evaluation of their toxicity and volatility. The first source of these properties is the safety data sheet (SDS) which all manufacturers are required to provide by federal regulation and in a prescribed format by the U.S. Occupational Safety and Health Administration (OSHA). An updated SDS is required to be provided by entities for each hazardous chemical annually to the SERC, LERC, and local fire department [OSHA 2012]. A searchable online database of these is available at Chemical Safety's website [Chemical Safety 2022]. The chemical hazards, toxicology, firefighting measures, and reactivity are all required sections of the SDS.

A planning team should meet for this operation and use the gathered data and expert knowledge to rank the HCs. The planning team should include a wide range of emergency management personnel, local government, and experts in the chemical field. A process of ranking can be subjective through expert opinion of the planning team involved or it can be made more objective by using a scoring system in key areas such as toxicity, inventory amounts, availability, volatility, and others. Score each category by its impact on the overall hazard level. All could be a 0-1 range, or a certain key category such as toxicity could be on a higher range (such as 0-2) to represent its relative importance to the other categories.

Unknown Chemicals

Unknown chemical risks, such as the intentional introduction of hazardous chemical agents by terrorists to create harm, panic, and disruption, will be addressed but will follow a modified approach due to more uncertainty in the EEI. Specifically, the chemical inventory uncertainties created by the lack of knowledge of the terrorist's chemical arsenal—and understanding the attacker goals—are critical gaps to creating the chemical risk profile. Plausible identification of the terrorist's chemical arsenal and attack goals can be identified through collaboration/interaction with the law enforcement intelligence community and/or chemical characterization conducted during incident through toxic syndrome (toxidrome) analysis.

Unknown chemicals are unidentified known chemicals at the time of the accident. The task is to create a risk profile of unknown chemicals by listing the health effects, or toxidromes, of the known chemicals. Information should be used from national, state, and local law enforcement that may help identify the unknown chemicals based on terrorist threat (Section 2.7). A good reference for toxidromes of unknown HC is compiled on the Chemical Hazards Emergency Medical Management (CHEMM) website [CHEMM 2021a]. A list of physiological effects of chemicals and other HC includes effects on:

- Pupils/vision
- Vital signs
- Skin
- Gastrointestinal tract
- Mucous membranes
- Urinary tract
- Mental confusion or delirium
- Muscles.

The defined symptoms for medical first responders are listed in 11 categories including irritant/corrosive toxidrome for inhalation or ingestion, knockdown agents in cellular or simple asphyxiants toxidromes, and convulsant toxidromes. An example of a toxidrome card provided by CHEMM [CHEMM 2021b] is shown for knockdown agents—simple asphyxiants in Figure 9. The card further lists examples of these agents such as cyanide, sodium azide, carbon monoxide, aniline, arsine, and nitrogen.

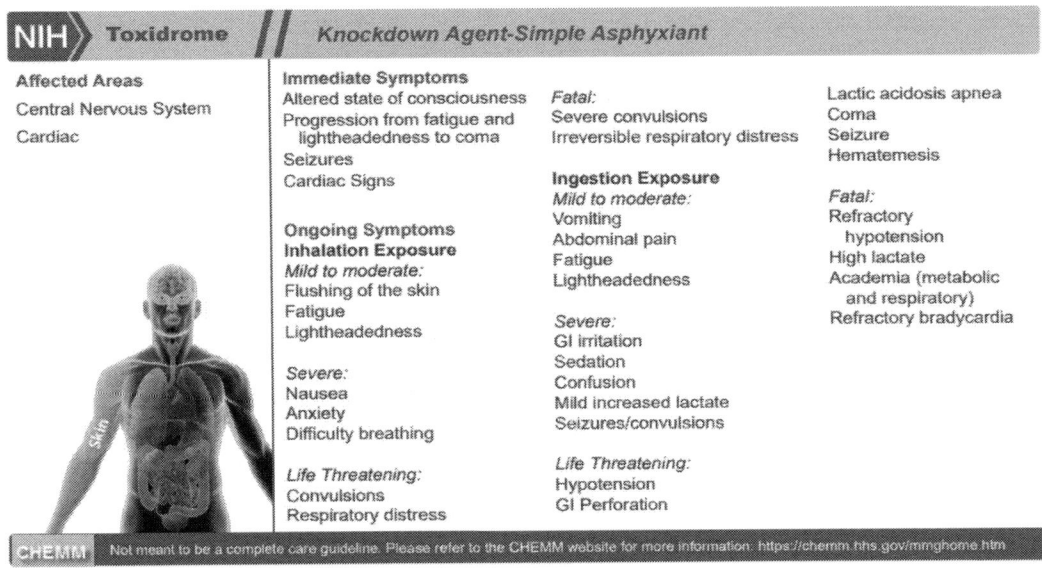

Figure 9. Toxidrome card example from CHEMM website (CHEMM 2021b).

Use the toxidromes to create an unknown chemical list to build and rank scenarios for planning alongside the known chemicals.

Determine Commodity Flows of Hazardous Chemicals

Transient chemical inventories pose an equally high risk compared to the static inventory used and/or produced in the community. The process of transportation exposes chemicals to traffic accidents on city streets, highways, railways, and waterways moving through the community. There is not a readily available source for assessing the inventory of transient chemicals. Even though OSHA and the U.S. Department of Transportation (DOT) [OSHA 2013] require placards and pictograms be placed on transportation vehicles and railcars the only records kept of the HC and their quantities are with shipping companies. Quantifying HC commodity flows through a community requires some research and cooperation from the shipping companies. HC commodity flows include:

- Railway resource scheduling and hazardous chemical inventory
- Highway/freight resource scheduling and hazardous chemical inventory

- Waterway resource scheduling and hazardous chemical inventory

- Pipeline resource scheduling and hazardous chemical inventory.

The U.S. Surface Transportation Board lists railway traffic including the number of tanker cars but only chemicals tracked are crude oil and ethanol [STB 2022].

The U.S. Federal Motor Carrier Safety Administration requires carriers and shippers to register as hazardous materials (HM) transporters and many states require further certification. Still, there is not a comprehensive database for how much HM moves through a highway corridor. The best potential resources for data are the local shipping companies.

There is less publicly available information about chemical commodity flow through seaports. The U.S. Department of Transportation Maritime Administration only lists specific regulations to follow. The amount and type of HC moving through pipelines is only available from the company owning the pipeline. Some of this information can overlap with the inventory of the production company.

A commodity Flow Study (CFS) provides a process for a community to evaluate and understand the movement of hazardous chemicals through different transportation routes in various regions of a locality. The conduct of a CFS should match the community's goals and available resources. Sources of data will be different from those used to identify chemical inventories for fixed facilities and may not be readily available. However, rough estimates of the movement of hazardous materials can provide important information for the determination of high priority risks. The DOT recommends the publication Guidebook for Conducting Local Hazardous Material Commodity Flow Studies [National Academy of Sciences, Engineering, and Medicine 2011]. This is a stepwise strategy that details various components of a comprehensive approach.

Population Analysis

Population information for a community should be compiled that includes population density and social vulnerability. Additionally, other aspects of population dynamics need to be considered for work/school hours/events to evaluate potential impacts to population based on time-of-day, week, and event. Areas of focus are:

- Resident population densities in a community

- Workday population densities in a community

- Vulnerable and less mobile population densities in a community such as elderly, hospitalized, and homeless communities

- Time-dependency of the population densities.

This step involves reviewing a map of the population densities of where people live, focusing on the densities of the population on a typical workday. These consist of locations such as downtown businesses, manufacturing facilities, schools, and public buildings. Analysis maps the densities of the population that cannot readily evacuate or have inadequate shelter-in-place options. This includes hospitals, nursing and long-term care facilities, and homeless/transient populations. Lastly, population densities of the entire population should be determined by day and hour.

Support for population analysis beyond local records can be found at the IPUMS (originally Integrated Public Use Microdata Series) National Historical Geographic Information System (NHGIS) [IPUMS 2022]. This is a free public database for determining population data.

Critical Sites and Venues

Important community sites/venues and critical population concerns related to dense public gatherings and critical functions/sites in the community should be identified. Population density is the main contributor to potential casualties and worst-case planning should include venues that congregate people together, impair the ability of emergency responders to react to the incident, provide a means to expand the incident's impact (such as a contaminated water supply), or affect a low-income or vulnerable population that would have trouble sheltering in place and/or evacuating efficiently.

List of critical sites and venues

A list of critical sites and venues should be developed using the criteria listed in Section 2.5. Examples of sites and venues that should be considered are listed in Table 1.

Table 1. Critical sites and venues.

Site/Venue	Reason for Inclusion			
	Population Density	Emergency Response Capability	Expanded Impact	Vulnerable Population
Hospital	X	X		X
Fire Stations, Law Enforcement		X		
Critical Infrastructure (water, power, etc...)			X	
Schools	X			X
Municipal Buildings	X			
Nursing and Long-Term Care Facilities				X
Holiday Venues	X			
Sports Venues	X			
Convention Centers	X			

Prioritization of the sites and venues should be subsequently conducted that is initially based on the role in an emergency response. For example, hospitals are critical to the treatment of persons impacted by a chemical release. A treatment facility located within the plume of a chemical release could be limited in their ability to provide appropriate care due to issues of contamination. Alternatively, municipal buildings, integral to the coordination and management of a response, would likely have continuity of operations (COOP) protocols in place that would provide them with an ability to quickly re-locate and continue the emergency operations coordination. Hospitals do not have the ability to re-locate operations resulting from the need for in-house treatment resources and an existing vulnerable patient population.

Enabling Factors

Enabling factors include what data is available to a community to support emergency management planning in terms of meteorological and disaster data, previous accidents/incidents, understanding HC system vulnerability, and unknown accidental or intentional HC release.

Weather history and current data sources

All aspects of weather can affect a HC release. Wind can move the HC over a wider population, precipitation can help or hinder the HC release, and temperature can have an impact both on the frequency of accidents occurring and on the severity of the accident. Near term risk assessment can use weather predictions from numerous sources. Most sources rely on the U.S. National Oceanic and Atmospherics' (NOAA) National Weather Service (NWS) [NOAA 2022].

Wind data

Wind is a major factor that affects a chemical release. When a chemical release occurs, local weather can be monitored through the nearest NWS site online. Planning for chemical release response needs weather data that predicts what winds will be like at the accident site on the day of the accident scenario. A wind rose is a good resource for planning where winds will likely move the chemical plume on any given day. Information includes wind speed (average and percentage of category), direction, and frequency of calm days (Figure 10). The Midwest Regional Climate Center (MRCC) at Purdue University, in cooperation with NOAA has a database of wind roses and other climate information on its cli-MATE database application [MRCC 2022] which lists temperature, wind, and precipitation records for all National Weather Service stations (see Figure 10). Cli-MATE requires a free registration. A wind rose dataset also resides at the US Department of Agriculture's Natural Resources Conservation Service National Water and Climate Center [USDA NRCS 2022]. This dataset provides prevailing winds for each month for selected cities.

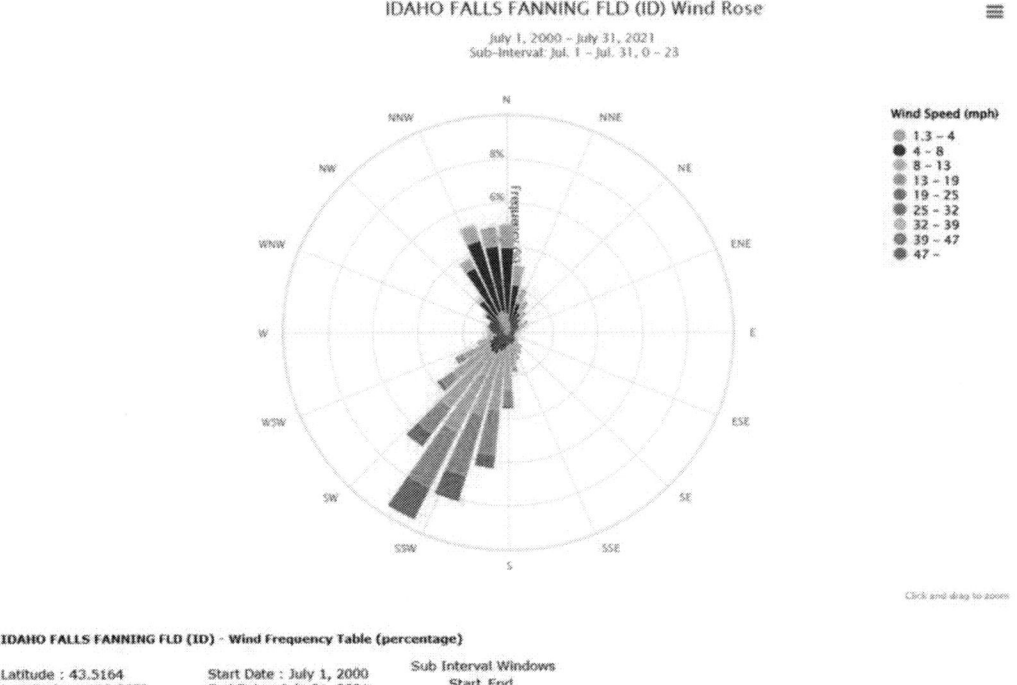

Figure 10. cli-MATE wind rose example from Purdue University and NOAA database (MRCC 2022).

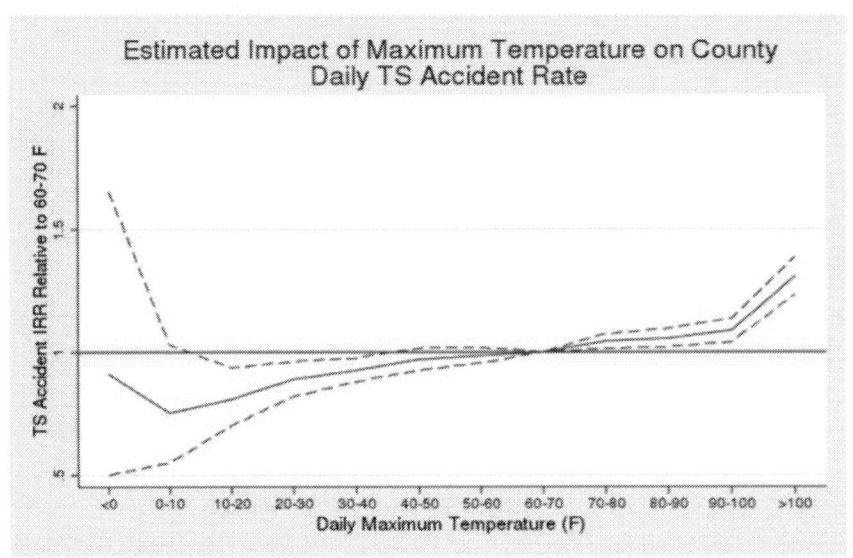

Figure 11. Temperature effects on temperature sensitive industrial accidents (Page, L., and S. Sheppard 2019).

Air temperature data

Human factor heat-related accidents are a documented hazard. According to research performed by Massachusetts Institute of Technology and Williams College [Page, L., and S. Sheppard 2019] a 105°F day causes 38.6% more workplace accidents in "temperature sensitive industries" than a day between 65-70°F (see Figure 11). Chemical industrial workplaces tend to be outdoor activities for fueling, transportation, and many production site tasks. Generally, accident frequencies are less in colder temperatures. Further, some hazardous chemicals are affected by temperatures, for example see the chemical SDSs [Chemical Safety 2022] for the effects of ambient air temperature and surface temperatures on chemicals of interest.

There are many resources for determining the expected temperatures for planning purposes. Most resources rely on the NWS.

Precipitation data

Precipitation can have positive or negative effects on chemicals of interest. Consult the chemical of interest' SDS. Precipitation averages for month/day can be found through many resources. Again, most of these resources rely on data from the NWS.

Historical information of disasters or critical infrastructure failures in the community

News sources and research can be used to gather information on prior CI systems failures and other disasters in the community of interest and in any representative community with similar infrastructure. These events directly affect the potential for HC release. A database listing should include:

- Location

- Date

- CI type

- Specific causes and effects.

The information gathered to apply to chemical accident scenarios can be used, such as:

- Transportation accidents that could happen to a chemical carrier

- Failure of CI that could affect chemical storage, transportation, or use

Lessons learned and applied from the research can be documented. This includes what physical structures or processes changes were proposed following the CI failure that were meant to prevent a re-occurrence. The HC release planning and response team should answer the following questions:

- Were these changes implemented?

- What is the current state of these structures and processes and how does that affect HC release?

- What physical structures or processes changes do the chemical incident planners and responders recommend safeguarding chemicals from these accidents?

Historical information of HC accidents and near misses in the community

News sources and research can be used including national and international databases of industrial accidents to list HC accidents that are applicable to the community of interest. The U.S. Chemical Safety and Hazard Investigation Safety Board (CSB) lists current and closed investigations reports on their website [CSB 2022]. Search by city, state, or incident type.

France's Bureau for Analysis of Industrial Risks and Pollutions (BARPI) from Analysis Research and Information on Accidents (ARIA) international industrial accident database has nearly all the U.S. accidents and a vast array of international accidents as well [BARPI ARIA 2022]. Type in the keyword in the search box on the top of the page, for instance "anhydrous ammonia" to find reports (see Figure 12).

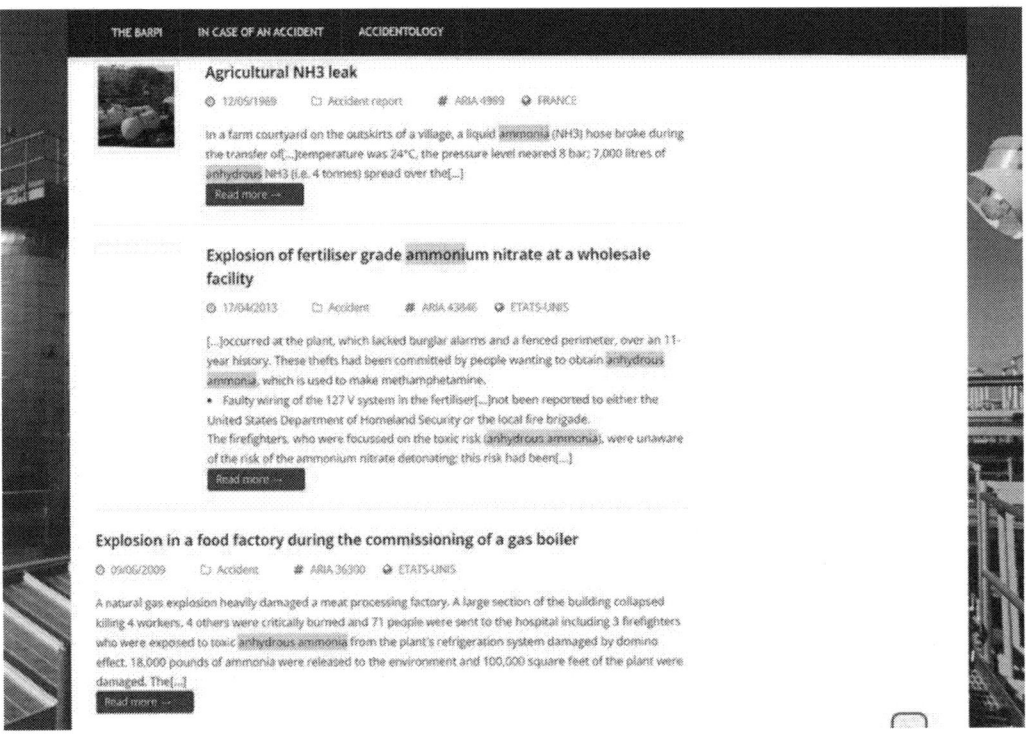

Figure 12. BARPI ARIA database example output for anhydrous ammonia (BARPI ARIA 2022).

Terrorism Precursors

Hazardous chemical releases and biological toxins have been used in high-profile terrorist acts such as the 1994-95 Tokyo subway sarin gas attacks, the 2001 US anthrax attacks, and the 2014 Chicago chlorine attack at a convention. How can local communities help prevent such occurrences? A common thought is that we rely on our state, federal, and even international law enforcement agencies to monitor the bad guys and let us know when they may be up to something. While this is true, it is important to recognize these larger well-equipped organizations rely on local community citizens and first responders to report something that may be suspicious.

Every state and most territories have set up "fusion centers" in cooperation with DHS. The fusion center's purpose is to create two-way communication and cooperation for threat-related information between DHS and state, local, tribal, territorial, federal, and private sector partners. One of the functions of the fusion center is to help quantify chemical threat levels primarily through a "see something, say something" approach.

To use fusion centers to their full advantage it is helpful to know what should be looked for. Terrorism, by design, has several precursors that law enforcement has found useful to monitor. What to watch out for in your community starts with the basics presented by DHS in their informational graphic on "see something, say something" (Figure 13) [DHS 2022].

The International Criminal Police Organization (INTERPOL) offers a dataset of profiles of people and entities associated with chemical, biological and improvised explosive device activities to member countries, including the United States, through Project Watchmaker [INTERPOL 2022a]. This is information that would usually be passed from INTERPOL to the Federal Bureau of Investigations (FBI) or DHS to the state fusion center to the community planners/first responders. INTERPOL also offers training videos such as one encouraging first responders to not operate in professional silos and cooperate through their Project Litmus video [INTERPOL 2022b].

Further training in spotting terrorist precursors is offered by the Federal Bureau of Investigations through videos on spotting suspicious retail sales, chemical laboratory security awareness, and others. Visit the FBI website at http://www.fbi.gov and search for "spotting terrorist precursors" for an extensive list. New Mexico Technical College offers DHS sponsored training on prevention and response to terrorist activities for first responders [NMT 2022].

These courses are not specifically directed at a chemical terrorist threat, but many common lessons can be learned.

Figure 13. DHS infographic on recognizing the signs of potential imminent terrorist activity (DHS 2022).

Results: Scenarios Leading to a Release

A key modeling step in risk assessment is the identification of chemical accident scenarios. This modeling of "what can go wrong?" follows the systematic identification of hazards, associated IE, and grouping of individual causes into like categories. In general, accident scenarios are the result of an upset condition, or IE, and the consequences following the upset condition. Note that IEs may lead directly to undesirable outcomes (a chemical release) or may require additional mitigative failures prior to reaching a negative outcome.

For the chemical hazard under consideration, the IEs that are evaluated are those that potentially may result in the undesired outcome of interest. Previously, we described the concept of a "scenario" showing how adverse consequences leading to a spectrum of consequences might occur. In this scenario representation, hazards may impinge on a community in different ways:

- They may provide enabling events (i.e., conditions that provide the opportunity to challenge mitigative systems, potentially leading to an accident)

- They may affect the occurrence of IEs
- They may defeat mitigating systems
- They may fail to reduce the consequences of mitigating system failures.

Communities are safe because, while hazards do exist, IEs do not occur very often. In addition, even if a chemical hazard does lead to an IE, mitigative systems (e.g., protective systems, building containment, and evacuation) typically work to some degree. Nonetheless, it is the goal of a risk assessment to assess hazards and provide a systematic approach to understanding, managing, and mitigating hazards that may exist in a community.

Identification and Ranking of Initiating Events

An exhaustive list of IEs should be considered when beginning scenario assessment. The planning team should brainstorm to identify as many IEs as possible that could cause the undesired outcome (a HC release), as follows:

- Types of accidents that can generate an IE list:
    - Static leakage
        - Inventory storage, tank configuration, HC present in equipment, etc.
    - Handling, fueling
    - Transportation
        - Highway, rail, air, pipeline, and waterway
    - Intentional release
        - Sprayed in air, added to water, detonated, etc.

Another approach to identifying IEs is through an Accident Event Sequence Diagram (AESD). The AESD is essentially a flowchart, with paths leading to different outcomes. A good example of an AESD is provided by the Canadian government's Road Safety at Work website [RSW 2019] (Figure 14). Note that when building an AESD it is good to start at the event and work backwards to define the IEs that caused the event. When all paths are complete there will be a list of IEs that can happen individually or must happen in sequence to consider.

Once an exhaustive list of IEs is created, there are several things to consider when determining what IEs to include in scenarios leading to a release. The most critical factors to consider are the frequency estimation and the severity of the HC release.

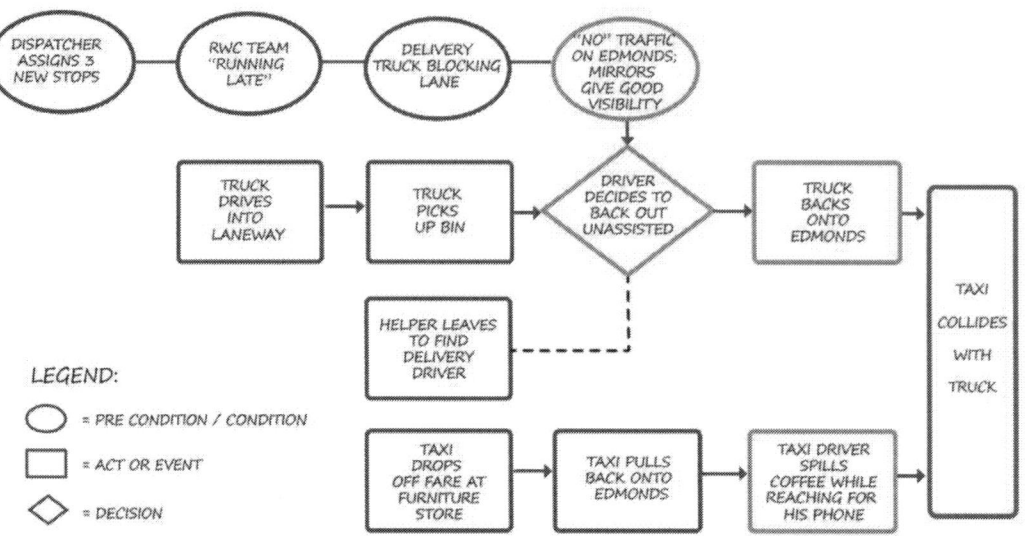

Figure 14. Accident Event Sequence Diagram example using a traffic accident (RSW 2019).

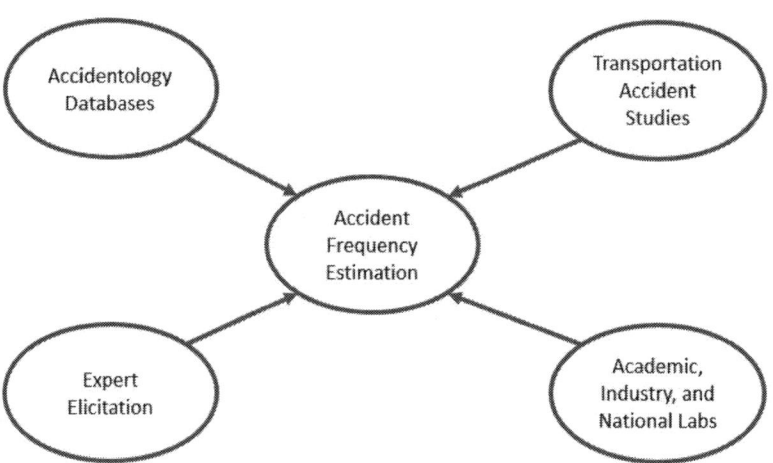

Figure 15. Inputs to frequency estimation.

Likelihood of a Release

The likelihood (frequency estimation) of the occurrence of an IE should be as accurate as possible. Frequency estimations can come from various sources. When multiple events must occur for the accident to happen, such as in Figure 14, the overall frequency must be determined. A planning engineer can apply a quantitative (numeric) approach to calculate or estimate a frequency estimation from industrial accident databases, transportation accident databases, academic, industry, or national laboratory reports and papers. The planning team can also use a more qualitative (non-numeric) approach by performing expert elicitation of the community planning team which includes first responders, and possibly other subject matter experts. An IE might have inputs from one or more of the quantitative or qualitative sources (Figure 15). While there are mathematical ways such as Bayesian analysis [Kelly, Dana, and Curtis Smith 2014] to weigh the impact of the inputs of all sources, it is not an absolute necessity. The planning team can use expert opinion on all gathered information to rank the frequency of the IE. The degree of quantitative versus qualitative analysis is up to the planning team.

Consequences of a Release

The most prominent component of consequence of a HC release is the severity (quantity, toxicity, and volatility). The severity of the HC release is determined and should be available from the chemical inventory step, which should provide a ranked order of the HCs available for use. The planning team should review the severity of the HC released for this step and account for the following to determine the consequence of the scenario:

- Location

- Weather

- Mitigations available

- Long-term effects

- Exposure banding.

The process of exposure banding helps to expand on the original toxicity/volatility assessment used in the severity assessment. Exposure banding as introduced by National Institute for Occupational Safety and Health [Lentz, T J, C Whittaker, L T McKernan, and M Seaton 2019, 132] is expanded somewhat to allow for unknown chemicals that would potentially reside outside of the occupational exposure bands but be defined by the effects noted in the exposed individuals (toxidromes). There are three classes of HC: known with documented upper exposure limit (UEL) and lower explosive concentration limit (LEL), known where UEL and/or LEL are not documented, and unknown HC where only the toxidromes are observed. All these HCs are defined in the chemical inventory step. The planning team will use the scenario built to determine the consequences based on the HC, the severity in the scenario, and the ability to mitigate the event based on resources, location, weather, short- and long-term effects (Figure 16).

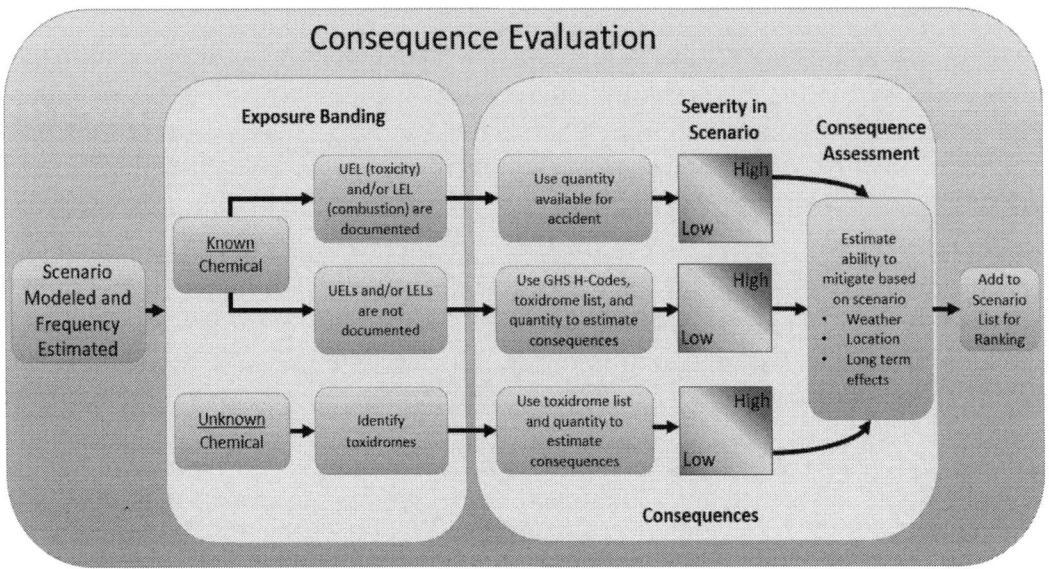

Figure 16. Consequence evaluation flow chart.

Ranking Scenarios

It is important to rank scenarios because it is unlikely that all scenarios can be assessed and planned for. By ranking the scenarios, the team is planning for those of highest risk. The planning team will use the frequencies of occurrence and the consequence evaluation when ranking the scenarios. Ideally, the highest ranked scenarios would be addressed through tabletop exercises with planning and emergency response teams. Lower risk scenarios not chosen for ranking, should they occur, would benefit from tabletop exercises performed on the higher risk scenarios.

Risk management, while not a focus of this paper, is the logical conclusion of the scenario evaluation. Those hazardous situations that have high risk should be considered for prevention (where possible) and then a community-level mitigation that is commensurate with the risk level.

Conclusion

The DRICAT framework for community HC and chemical risk assessment was presented. DRICAT is a framework and is not intended to mandate specific ways to perform each of the three steps, nor is it meant to recommend any specific software or other tool. For instance, there are many methods to elicit expert opinion and it is up to the planning team to decide which is best for them. There are many sources of weather data, chemical data sheets, accidentology information, etc. The framework draws from publicly available sources referenced throughout the article.

DRICAT is focused on planning for HC accidents. The technique quantitatively and/or qualitatively determines the frequencies and consequences of the accidents to determine what are the most important scenarios to further investigate through tabletop exercises with community planning and emergency response teams. The intent of DRICAT is to increase community awareness of the location's chemical risk inventory, vulnerable sites and populations, and the publicly available resources available for assessing the risk of a HC accidental or deliberate release.

Acknowledgements and Funding

Research reported in this publication was supported by the Countering Weapons of Mass Destruction Office of the U.S. Department of Homeland Security under award number 70RWMD21K00000011. The content is solely the responsibility of the authors and does not necessarily represent the official views of U.S. Department of Homeland Security. Neither the U.S. Government, nor any agency thereof, nor any of their employees makes any warranty, express or implied, or assumes any legal liability or responsibility for the accuracy, completeness, or usefulness of any information, apparatus, product, or process disclosed, or represents that its use would not infringe privately owned rights.

Acronyms and Abbreviations

AESD	accident event sequence diagram
ARIA	Research and Information on Accidents
BARPI	France's Bureau for Analysis of Industrial Risks and Pollutions Analysis
CFS	Commodity Flow Study
CHEMM	Chemical Hazards Emergency Medical Management
CI	critical infrastructure
COOP	continuity of operations
CSB	The U.S. Chemical Safety and Hazard Investigation Safety Board
CSDS	chemical safety data sheet
DHS	U.S. Department of Homeland Security
DOT	U.S. Department of Transportation
DRICAT	Data and Risk-Informed Chemical Assessment Technique
EEI	Essential Elements of Information
EPCRA	Emergency Planning and Community Right-to-Know Act
FBI	Federal Bureau of Investigations
FEMA	Federal Emergency Management Agency

HC	hazardous chemical
HID	hazard identification diagram
HM	hazardous material
IE	initiating event
INTERPOL	The International Criminal Police Organization
LEL	lower explosive concentration limit
LEPC	Local Emergency Planning Committee
MLD	master logic diagram
MRCC	Midwest Regional Climate Center at Purdue University
NOAA	U.S. National Oceanic and Atmospheric Administration
NWS	U.S. National Weather Service
SERC	State Emergency Response Commission
UEL	upper exposure limit

Capsule Author Bios

Curtis L. Smith, Ph.D. is the Director for the Idaho National Laboratory Nuclear Safety and Regulatory Research Division. Prior to taking a leadership role as Division Director, he led several risk-informed activities including: the Risk-Informed Systems Analysis (RISA) Pathway under the DOE Light Water Reactor Sustainability Program; the NRC's SAPHIRE risk analysis software development; the NASA Safety Mission Success project at INL; the Risk Integration and Uncertainty Working Group for the NASA Interagency Nuclear Safety Review Panel for both the Mars 2020 and Mars Science Laboratory missions. Dr. Smith has been at INL for 31 years and has published over 275 papers, books, and reports on risk and reliability theory and applications. He holds a Ph.D. in nuclear engineering from MIT.

Kurt G. Vedros is a lead risk assessment engineer for Idaho National Laboratory's Nuclear Science and Technology division Reliability, Risk, and Resilience Sciences department. He has 26 years of experience in reliability and risk engineering with an emphasis on electrical, chemical, and nuclear power plant-based systems. He has applied probabilistic risk assessment to high-risk electrical grids for the Department of Defense and other customers. His current research interests include Bayesian inference applications for parameter estimations and environmental risk, artificial intelligence and machine learning, and novel electrical grid risk metrics. Kurt received his Associates of Electrical Engineering Technology from the Community College of the Air Force in 1992, completed Reliability Engineering Institutes I & II at University of Arizona, USA in 1996, and he received a Bachelor of Science degree in Nuclear Engineering from Idaho State University, Pocatello, Idaho, USA in 2007.

Kenneth F. Martinez is a Certified Industrial Hygienist and Environmental Engineer with experience in leading and conducting large-scale research; managing programs in occupational safety and health, and emergency response; and creating and teaching professional development courses. He brings 33 years of CDC expertise in the area of hazardous agent exposure characterization and mitigation control practices in the manufacturing and healthcare industry. Ken is a recognized subject matter expert in biological agents including infectious disease and bioterrorism agents. Over a nine-year period for CDC, he served in numerous emergency response field leadership roles including the World Trade Center collapse, anthrax, SARS, multi-drug resistant TB, hurricane emergency responses (Katrina and Wilma) and the Deep-Water Horizon oil spill. As a subject matter expert on biological threat agents, Ken provided counsel and leadership to senior management of other federal agencies including providing testimony before a congressional sub-committee on Capitol Hill, presenting before senior leaders of the OSTP, GAO, USPS, DHS, EPA, and FBI and serving on a CDC team tasked to provide responses and a strategic plan to the Pandemic and All-Hazards Preparedness Act (PAHPA). During 2009, the H1N1 pandemic, Ken coordinated and directed education through media communication to over 600 Cincinnati NIOSH staff. His research efforts in bioaerosols have resulted in 38 peer-reviewed journal articles and book chapters; 32 NIOSH technical reports; over 100 technical presentations at scientific meetings; and courses on bioaerosols targeted at the local, state, and national level. From 2013 to 2020, he supported DHS as a contractor for a national airborne biodetection system serving as a Jurisdictional Coordinator for two jurisdictions, as an Environmental Assessment (EA) team lead and as a Field Operations team lead. Subsequently, he provided contract support to FEMA CBRN and FDA. He currently serves as a Senior Critical Infrastructure Analyst for Idaho National Laboratory and concurrently operates as an executive officer for a non-profit created to provide support of COVID-19 pandemic response efforts nationally.

David G. Kuipers is a critical infrastructure cyber security engineer for Idaho National Laboratory's National and Homeland Security division. David received a Bachelor of Science in Technology from the University of Idaho, Moscow, Idaho, USA in 2002. He has 45 years' experience working in process systems working in nuclear systems in operations and engineering, and engineering design of industrial control systems (ICS) in nuclear, electrical, and water process applications. He developed a strong background in critical infrastructure systems and operations. David was the Program Manager for the INL National SCADA Test Bed program working with the largest electrical ICS vendors and industry operators, testing cybersecurity, and supporting vendor systems improvements to strengthen the US power grid. Most recent years he has been working directly with multi-sector entities to evaluate yber/physical systems-of-systems weaknesses to improve security posture.

References

BARPI ARIA. 2022. *The ARIA Database*. La référence du retour d'expérience sur accidents technologiques. https://www.aria.developpement-durable.gouv.fr/the-barpi/the-aria-database/?lang=en.

Chemical Safety. 2022. *Free Safety Data Sheet Search*. Chemical Safety. https://chemicalsafety.com/sds-search.

CHEMM. 2021a. *Toxic Syndromes/Toxidromes*. Chemical Hazards Emergency Medical Management. U.S. Department of Health and Human Services. https://chemm.hhs.gov/toxicsyndromes.htm.

CHEMM. 2021b. *CHEMM Toxidrome Cards*. Chemical Hazards Emergency Medical Management. U.S. Department of Health and Human Services. https://chemm.hhs.gov/toxidrome_cards.htm.

CISA. 2022. *Chemical Security*. Cybersecurity & Infrastructure Security Agency. https://www.cisa.gov/chemical-security.

CSB. 2022. *Chemical Accident Investigations*. U.S. Chemical Safety and Hazard Investigation Safety Board. https://www.csb.gov/investigations.

DHS. 2010. *DHS Risk Lexicon*. Washington, DC: Department of Homeland Security.

DHS. 2015. *Chemical Sector Specific Plan*. Washington, D.C.: U.S. Dept. of Homeland Security.

DHS. 2022. *Recognize the Signs*. U.S. Department of Homeland Security. https://www.dhs.gov/see-something-say-something/recognize-the-signs.

EPA. 1999. *How Safe Am I? Helping Communities Evaluate Chemical Risks*. Environmental Protection Agency.

EPA. 2022a. Protecting Communities from Chemical Accidents: Emergency Planning and Community Right-to-Know Act. Environmental Protection Agency. https://www.epa.gov/epcra/protecting-communities-chemical-accidents-emergency-planning-and-community-right-know-act.

EPA. 2022b. *EnviroAtlas.* Environmental Protection Agency. enviroatlas.epa.gov/enviroatlas/interactivemap.

FEMA. 2013. *National Incident Support Manual.* Washington, D.C.: U.S. Dept. of Homeland Security. Federal Emergency Management Agency.

INTERPOL. 2022a. *Project Watchmaker.* The International Criminal Police Organization – INTERPOL. https://www.interpol.int/en/Crimes/Terrorism/Chemical-and-Explosives-terrorism/Project-Watchmaker.

INTERPOL. 2022b. *Project Litmus.* The International Criminal Police Organization – INTERPOL. https://www.interpol.int/en/Crimes/Terrorism/Chemical-and-Explosives-terrorism/Project-Litmus.

IPUMS. 2022. *IPUMS NHGIS: Version 15.0.* IPUMS. https://www.ipums.org/projects/ipums-nhgis/d050.v15.0.

Kaplan, Stanley, and B. John Garrick. 1981. *On the Quantitative Definition of Risk.* Risk Analysis 1, no. 1 (1981): 11–27. https://doi.org/10.1111/j.1539-6924.1981.tb01350.x.

Kelly, Dana, and Curtis Smith. 2014. *Bayesian Inference for Probabilistic Risk Assessment: A Practitioner's Guidebook.* Springer London Ltd.

Lentz, T J, C Whittaker, L T McKernan, and M Seaton. 2019. *The NIOSH Occupational Exposure Banding Process for Chemical Risk Management.* Vol. 132. Cincinnati, OH.: Department of Health and Human Services, National Institute for Occupational Safety and Health, 2019. https://doi.org/10.26616/NIOSHPUB20

MRCC. 2022. *Cli-Mate: MRCC Application Tools Environment.* Midwest Regional Climate Center, Purdue University. https://mrcc.purdue.edu/CLIMATE.

National Academies of Sciences, Engineering, and Medicine. 2011. *Guidebook for Conducting Local Hazardous Materials Commodity Flow Studies.* Washington, D.C: The National Academies Press.

NMT. 2022. *New Mexico Technical University Counterterrorism First Responder Training.* DHS Sponsored First Responder Training - Respond. New Mexico Technical College. http://www.emrtc.nmt.edu/training.

NOAA. 2022. *National Weather Service.* U.S. Department of Commerce. https://www.weather.gov/.

OSHA. 2012. *Hazard Communication Standard: Safety Data Sheets.* U.S. Occupational Safety and Health Administration, Communication Brief 3514. https://www.osha.gov/sites/default/files/publications/OSHA3514.pdf.

OSHA. 2013. *Hazard Communication Standard: Labels and Pictograms.* U.S. Occupational Safety and Health Administration, Communication Brief 3636. https://www.osha.gov/sites/default/files/publications/OSHA3636.pdf.

Page, L., and S. Sheppard. 2019. *Heat Stress: Ambient Temperature and Workplace Accidents in the US.* Massachusetts Institute of Technology, Williams College, 2019.

RSW. 2019. *Step 3: Map the Sequence of Events: Road Safety at Work.* Road Safety at Work, Justice Institute of British Columbia. https://roadsafetyatwork.ca/toolkits/investigating-motor-vehicle-incidents/conducting-mvi-investigations/step-3-map-the-sequence-of-events/.

STB. 2022. *Rail Service Data.* Surface Transportation Board. https://www.stb.gov/reports-data/rail-service-data.

USDA NRCS. 2022. Wind Rose Resources. U.S. Department of Agriculture National Resources Conservation Service. https://www.nrcs.usda.gov/wps/portal/wcc/home/climateSupport/windRoseResources.

Related Titles from Westphalia Press

The Zelensky Method
by Grant Farred

Locating Russian's war within a global context, The Zelensky Method is unsparing in its critique of those nations, who have refused to condemn Russia's invasion and are doing everything they can to prevent economic sanctions from being imposed on the Kremlin.

China & Europe: The Turning Point
by David Baverez

In creating five fictitious conversations between Xi Jinping and five European experts, David Baverez, who lives and works in Hong Kong, offers up a totally new vision of the relationship between China and Europe.

Masonic Myths and Legends
by Pierre Mollier

Freemasonry is one of the few organizations whose teaching method is still based on symbols. It presents these symbols by inserting them into legends that are told to its members in initiation ceremonies. But its history itself has also given rise to a whole mythology.

Resistance: Reflections on Survival, Hope and Love
Poetry by William Morris, Photography by Jackie Malden

Resistance is a book of poems with photographs or a book of photographs with poems depending on your perspective. The book is comprised of three sections titled respectively: On Survival, On Hope, and On Love.

Bunker Diplomacy: An Arab-American in the U.S. Foreign Service
by Nabeel Khoury

After twenty-five years in the Foreign Service, Dr. Nabeel A. Khoury retired from the U.S. Department of State in 2013 with the rank of Minister Counselor. In his last overseas posting, Khoury served as deputy chief of mission at the U.S. embassy in Yemen (2004-2007).

Managing Challenges for the Flint Water Crisis
Edited by Toyna E. Thornton, Andrew D. Williams, Katherine M. Simon, Jennifer F. Sklarew

This edited volume examines several public management and intergovernmental failures, with particular attention on social, political, and financial impacts. Understanding disaster meaning, even causality, is essential to the problem-solving process.

Donald J. Trump, The 45th U.S. Presidency and Beyond International Perspectives
Editors: John Dixon and Max J. Skidmore

The reality is that throughout Trump's presidency, there was a clearly perceptible decline of his—and America's—global standing, which accelerated as an upshot of his mishandling of both the Corvid-19 pandemic and his 2020 presidential election loss.

Brought to Light: The Mysterious George Washington Masonic Cave
by Jason Williams, MD

The George Washington Masonic Cave near Charles Town, West Virginia, contains a signature carving of George Washington dated 1748. Although this inscription appears authentic, it has yet to be verified by historical accounts or scientific inquiry.

Abortion and Informed Common Sense
by Max J. Skidmore

The controversy over a woman's "right to choose," as opposed to the numerous "rights" that abortion opponents decide should be assumed to exist for "unborn children," has always struck me as incomplete. Two missing elements of the argument seems obvious, yet they remain almost completely overlooked.

The Athenian Year Primer: Attic Time-Reckoning and the Julian Calendar
by Christopher Planeaux

The ability to translate ancient Athenian calendar references into precise Julian-Gregorian dates will not only assist Ancient Historians and Classicists to date numerous historical events with much greater accuracy but also aid epigraphists in the restorations of numerous Attic inscriptions.

The Politics of Fiscal Responsibility: A Comparative Perspective
by Tonya E. Thornton and F. Stevens Redburn

Fiscal policy challenges following the Great Recession forced members of the Organisation for Economic Co-operation and Development (OECD) to implement a set of economic policies to manage public debt.

Growing Inequality: Bridging Complex Systems, Population Health, and Health Disparities
Editors: George A. Kaplan, Ana V. Diez Roux, Carl P. Simon, and Sandro Galea

Why is America's health is poorer than the health of other wealthy countries and why health inequities persist despite our efforts? In this book, researchers report on groundbreaking insights to simulate how these determinants come together to produce levels of population health and disparities and test new solutions.

Issues in Maritime Cyber Security
Edited by Dr. Joe DiRenzo III, Dr. Nicole K. Drumhiller, and Dr. Fred S. Roberts

The complexity of making MTS safe from cyber attack is daunting and the need for all stakeholders in both government (at all levels) and private industry to be involved in cyber security is more significant than ever as the use of the MTS continues to grow.

A Radical In The East
by S. Brent Morris, PhD

The papers presented here represent over twenty-five years of publications by S. Brent Morris. They explore his many questions about Freemasonry, usually dealing with origins of the Craft. A complex organization with a lengthy pedigree like Freemasonry has many basic foundational questions waiting to be answered, and that's what this book does: answers questions.

Contests of Initiative: Countering China's Gray Zone Strategy in the East and South China Seas
by Dr. Raymond Kuo

China is engaged in a widespread assertion of sovereignty in the South and East China Seas. It employs a "gray zone" strategy: using coercive but sub-conventional military power to drive off challengers and prevent escalation, while simultaneously seizing territory and asserting maritime control.

Frontline Diplomacy: A Memoir of a Foreign Service Officer in the Middle East
by William A. Rugh

In short vignettes, this book describes how American diplomats working in the Middle East dealt with a variety of challenges over the last decades of the 20th century. Each of the vignettes concludes with an insight about diplomatic practice derived from the experience.

Anti-Poverty Measures in America: Scientism and Other Obstacles
Editors, Max J. Skidmore and Biko Koenig

Anti-Poverty Measures in America brings together a remarkable collection of essays dealing with the inhibiting effects of scientism, an over-dependence on scientific methodology that is prevalent in the social sciences, and other obstacles to anti-poverty legislation.

Geopolitics of Outer Space: Global Security and Development
by Ilayda Aydin

A desire for increased security and rapid development is driving nation-states to engage in an intensifying competition for the unique assets of space. This book analyses the Chinese-American space discourse from the lenses of international relations theory, history and political psychology to explore these questions.

westphaliapress.org

Made in the USA
Middletown, DE
24 October 2022